SpringerBriefs in Physics

SpringerBriefs in Physics are a series of slim high-quality publications encompassing the entire spectrum of physics. Manuscripts for SpringerBriefs in Physics will be evaluated by Springer and by members of the Editorial Board. Proposals and other communication should be sent to your Publishing Editors at Springer.

Featuring compact volumes of 50 to 125 pages (approximately 20,000–45,000 words), Briefs are shorter than a conventional book but longer than a journal article. Thus, Briefs serve as timely, concise tools for students, researchers, and professionals.

Typical texts for publication might include:

- A snapshot review of the current state of a hot or emerging field
- A concise introduction to core concepts that students must understand in order to make independent contributions
- An extended research report giving more details and discussion than is possible in a conventional journal article
- A manual describing underlying principles and best practices for an experimental technique
- An essay exploring new ideas within physics, related philosophical issues, or broader topics such as science and society

Briefs allow authors to present their ideas and readers to absorb them with minimal time investment. Briefs will be published as part of Springer's eBook collection, with millions of users worldwide. In addition, they will be available, just like other books, for individual print and electronic purchase. Briefs are characterized by fast, global electronic dissemination, straightforward publishing agreements, easy-to-use manuscript preparation and formatting guidelines, and expedited production schedules. We aim for publication 8–12 weeks after acceptance.

Albert Schwarz

Quantum Mechanics and Quantum Field Theory from Algebraic and Geometric Viewpoints

Springer

Albert Schwarz
Department of Mathematics
University of California
Davis, CA, USA

ISSN 2191-5423 ISSN 2191-5431 (electronic)
SpringerBriefs in Physics
ISBN 978-3-031-67914-8 ISBN 978-3-031-67915-5 (eBook)
https://doi.org/10.1007/978-3-031-67915-5

This Springer imprint is published by the registered company Springer Nature Switzerland AG
The registered company address is: Gewerbestrasse 11, 6330 Cham, Switzerland

I dedicate this book to my family: Galina, Michael, Bella, Abby, Max, Liya, Serge, Jelena, and Amelia

Preface

In the conventional exposition of quantum mechanics, we inhabit Hilbert space and examine operators within this space. Self-adjoint operators are associated with physical quantities. Physicists predominantly use this methodology, however, it has its limitations. We will explore alternative approaches. Primarily, the algebraic approach, where the initial point is an algebra of observables, an associative algebra with involution ($*$-algebra), in which the self-adjoint elements are observables. This approach is nearly as old as quantum mechanics itself. In addition, we will discuss the geometric approach, where the initial point is a set of states. This idea was proposed in my papers several years ago and it is considerably more general than the algebraic approach. We demonstrate within the framework of this approach that quantum mechanics can be viewed as classical mechanics where our devices permit us to observe only a subset of physical quantities. Furthermore, we illustrate that in this manner we can construct a wide class of physical theories that generalize quantum mechanics.

Our exposition diverges from standard textbooks in quantum mechanics and quantum field theory in numerous ways.

In particular, we highlight that the emergence of probabilities in quantum theory can be elucidated from decoherence caused by adiabatic interaction with a random environment. We underscore that the concept of a particle is not primary in quantum theory. It is a secondary concept. If the theory is translation-invariant we define particles as elementary excitations of the ground state. Quasiparticles are elementary excitations of any translation-invariant state. We analyze the concept of scattering but we do not utilize the concept of a field and do not assume locality and Poincaré invariance. We will discuss not only the conventional scattering matrix (related to scattering cross-sections) but also the concept of an inclusive scattering matrix, which is closely related to the concept of inclusive scattering cross-sections. It is demonstrated that scattering matrices can be expressed in terms of Green's functions by the well-known formula belonging to Lehmann, Symanzik, and Zimmermann, and the inclusive scattering matrices can be expressed in terms of generalized Green's functions, which first appeared in non-equilibrium statistical physics in Keldysh formalism. As a concrete realization of the geometric approach, we describe the formalism of L-functionals where states are represented by non-linear functionals corresponding to positive functionals on Weyl and Clifford algebras (to states in the algebraic approach).

Chapter 1 of the book is devoted to basic concepts of quantum mechanics in conventional, algebraic, and geometric approaches. In Chap. 2, we consider particles, quasiparticles, and their scattering. One could argue that this chapter is dedicated to quantum field theory without the concept of a field. The first two chapters are based on the course I taught online in 2022 and those lectures were published as "Quantum Mechanics and Quantum Field Theory: Algebraic and Geometric Approaches" by I. Frolov and A. Schwarz (Universe 2023, 9, 337).

Disclaimer: All sections of Chaps. 1 and 2 except Sect. 2.8 and all sections of Appendix as well as Sect. 4.2 contain reprinted material from that source published under CC-BY-4.0 license.

Chapter 3 provides a concise exposition of relativistic theories from an algebraic perspective.

Chapter 4 is largely independent of the first chapters (to make the chapter more independent I repeated some definitions). A reader interested in new results that are proved in the book can start with this chapter referring to preceding chapters for details.

This chapter is devoted to deterministic physical theories considered from a geometric perspective. It is emphasized here that the geometric approach allows us to transcend quantum mechanics. It leads to a large class of theories that share many characteristics with quantum theory. (They are deterministic, but the analog of decoherence leads to probabilities in a random environment). The most intriguing new result is the assertion that the infrared divergences of quantum electrodynamics do not appear in the formalism of L-functionals described in this chapter.

The present book should be accessible to students familiar with classical mechanics and some basic mathematical notions (Hilbert space, self-adjoint operator, Banach space, topological space). Some necessary mathematical notions are explained in the appendix. On the other side, it contains many recent results that can be interesting to an accomplished physicist or mathematician.

I did not try to make the exposition completely rigorous (a mathematician can find more rigorous definitions in Appendix). However, it is easy to give mathematical proofs of the main theorems of scattering theory (and in many other cases it is not hard to give rigorous proofs imposing some additional conditions).

Bures-sur-Yvette, France Albert Schwarz
May 2024

Acknowledgments I am indebted to N. Arkani-Hamed, T. Damour, A. Goncharov, A. Gorsky, I. Frolov, A. Kapustin, M. Katsnelson, Z. Komargodsky, M. Kontsevich, J. Parra Martinez, M. Movshev, B. Nachtergaele, N. Nekrasov, A. Polyakov, A. Rosly, A. Smilga, J. Trnka, A. Vainshtein, M. Vasiliev, A. Waldron, O. Zaboronski for useful discussions. My special thanks to O. Zaboronski for the help with editing.

The original version of the book has been revised. A correction to this book can be found at https://doi.org/10.1007/978-3-031-67915-5_5

Contents

Chapter 1
Quantum Theory in Algebraic and Geometric Approaches

The algebraic approach to quantum theory was suggested very early mostly by von Neumann and Jordan [13, 19]. Later it was developed in numerous papers (usually for relativistic local theories). The geometric approach to quantum theory was suggested recently in papers [26, 27]. The derivation of decoherence based on consideration of adiabatic random perturbations follows [22] (see also later papers [1, 20]) A generalization of this derivation in geometric approach was given in [27].

The paper [10] **based on my lectures contains a review of algebraic and geometric approaches to quantum theory. This chapter mostly follows the same presentation. All sections include reprinted material from that review, which is openly available.**

1.1 Traditional Quantum Mechanics

We review some fundamental concepts of quantum mechanics. For our purposes, a state in quantum mechanics is represented by a density matrix. A density matrix is a self-adjoint operator K in Hilbert space $\mathcal{H}$, which is positive definite (more precisely, positive semi-definite) and has a trace equal to one: $\operatorname{Tr} K = 1$. The set of density matrices forms a convex set, and its extreme points are referred to as pure states. These pure states correspond to vectors in Hilbert space. Each normalized vector Ψ corresponds to a density matrix, defined as an orthogonal projection onto this vector:

$$K_{\Psi}(x) = \langle x, \Psi \rangle \Psi.$$

(For any non-zero vector, the density matrix is defined as $K_{\frac{\Psi}{||\Psi||}}$).

A. Schwarz, *Quantum Mechanics and Quantum Field Theory in Algebraic and Geometric Approaches*, SpringerBriefs in Physics,
https://doi.org/10.1007/978-3-031-67915-5_1

It should be noted that if two non-zero vectors are proportional ($\Psi' = \lambda\Psi$), then the corresponding density matrices are identical: $K_{\Psi'} = K_{\Psi}$.

The density matrix K has a basis of eigenvectors e_i with non-negative eigenvalues p_i that sum to 1. We can say that each of these vectors corresponds to a pure state, and the density matrix is a mixture of these pure states. To verify this, we use a representation in which the matrix K is diagonal, then the diagonal elements are equal to p_i. Since the trace equals 1, their sum equals one: $\sum p_i = 1$. Since the matrix is positive definite, the diagonal elements are non-negative: $p_i \geq 0$. Consequently, we can say that the density matrix is a mixture of pure states with probabilities p_i.

In standard quantum mechanics textbooks, this is explained in reverse order, starting with pure states. Density matrices are defined as mixed states.

We discussed one method of representing the density matrix as a mixture of pure states. There are infinitely many ways to do this.

A physical quantity is associated with a self-adjoint operator in $\mathcal{H}$. The expectation value $\bar{A}$ of a physical quantity $\hat{A}$ in the state described by the density matrix K is given by the formula $\bar{A} = \mathrm{Tr}\hat{A}K$.

Every self-adjoint operator defines a one-parameter group of unitary operators $e^{-i\hat{A}t}$. This group acts on states (on density matrices) transforming a density matrix K into $K(t) = e^{i\hat{A}t}Ke^{-i\hat{A}t}$.

In quantum mechanics, as in all theories we consider, we should have an evolution operator, an operator $U(t)$ transforming the state K at time 0 into the state $K(t) = U(t)K$ (the state at time t). The operators $U(t)$ obey $U(t+t') = U(t)U(t')$ (they form a one-parameter group). They can be written in the form $U(t)K = e^{i\hat{H}t}Ke^{-i\hat{H}t}$ where $\hat{H}$ is a physical quantity called the Hamiltonian (or energy operator). In principle, every self-adjoint operator can serve as a Hamiltonian; the choice of Hamiltonian specifies the physical system we are working with. The state $K(t)$ satisfies the equation

$$\frac{dK}{dt} = i(\hat{H}K - K\hat{H}) \tag{1.1}$$

(the equation of motion or Schrödinger equation). For pure states represented by vectors in Hilbert space, this equation takes the form

$$i\frac{d\Psi}{dt} = \hat{H}\Psi.$$

We are working in the Schrödinger picture where states are time-dependent. One can work in the Heisenberg picture where operators depend on time but the states do not. The Heisenberg operators $\hat{A}(t)$ are defined by the formula $\hat{A}(t) = e^{-i\hat{H}t}\hat{A}e^{i\hat{H}t}$; they obey the equation

$$\frac{dA}{dt} = i(\hat{A}\hat{H} - \hat{H}\hat{A})$$

(Heisenberg equation).

Note that in the above formulas, we used the system of units where the Planck constant $\hbar = 1$. It is convenient to restore $\hbar$ in these formulas when we consider the relation between quantum and classical mechanics; for example, the Heisenberg equation takes the form

$$\hbar\frac{dA}{dt} = i(\hat{A}\hat{H} - \hat{H}\hat{A}). \tag{1.2}$$

In the limit $\hbar \to 0$ quantum mechanics gives the same results as classical mechanics; in particular, the Heisenberg equation gives in this limit Hamiltonian equations of motion (see Sect. 1.5 for details).

1.2 Geometric and Algebraic Approaches to Quantum Theory

In the geometric approach, the initial point is the set of states. We postulate that the set of states is a bounded closed convex subset of a topological linear space. These assumptions are adequate to develop a substantial theory.

In the algebraic approach, the initial point is the algebra of observables $\mathcal{A}$. (See Appendix A.1 for a review of the fundamental algebraic notions used.) More specifically, we are interested in associative unital algebras equipped with an involution (such an algebra is called a $*$-algebra). A typical example of a $*$-algebra (the basic one for us) is the algebra of bounded operators on a Hilbert space. In this algebra, there is a bijection $A \to A^*$, which maps an operator to its adjoint. It is an involution, implying that the square of the $*$-map is an identity: $A^{**} = A$. Additionally, if we take the adjoint operator of a product, it will be the product again, but in the reverse order: $(AB)^* = B^*A^*$. Furthermore, the involution $*$ is antilinear, $(\alpha A + \beta B)^* = \bar{\alpha}A^* + \bar{\beta}B^*$. In the operator algebra, these are simple provable properties, but for arbitrary $*$-algebras, these are axioms.

We always assume that there is some topology in the algebra in which all operations are continuous. We usually won't discuss these topologies, first, because it is time-consuming, and second, because different topologies can be equally sensible. Sometimes it is necessary to have a norm in which the algebra is a Banach space, then one requires the inequality $||AB|| \leq ||A|| \cdot ||B||$ to hold (this is the definition of a Banach algebra). Sometimes we impose additional conditions, such as that the algebra is a C^*-algebra (this means that the norm of the product A^*A is equal to the square of the norm of the operator A, that is, $||A^*A|| = ||A||^2$). When considering a homomorphism or an automorphism of a $*$-algebra, we always assume that it is continuous and agrees with the involution.

If there is an algebra with involution, the self-adjoint elements ($A = A^*$) correspond to physical quantities. Self-adjoint elements themselves do not form a subalgebra (the product of self-adjoint elements is not necessarily a self-adjoint element). This disadvantage of the algebraic approach was noticed as early as the 1930s; this led to the notion of Jordan algebras. Jordan noticed that although the product of self-adjoint elements is not self-adjoint, the anticommutator $A \circ B := AB + BA$, where A and B are self-adjoint, is again self-adjoint. He axiomatized this operation.

The theory of Jordan algebras was constructed in the 1930s, mainly in the famous work of Jordan et al. [14]. But although Jordan algebra is a very beautiful and really useful object in many parts of mathematics, it has not yet had much use in physics. Now it has naturally appeared in the geometric approach [30] and may come back to physics again.

If we start from a $*$-algebra (an associative algebra with involution), we can define the notion of state: a state is a linear functional ω on algebra A that satisfies the non-negativity condition if restricted to the elements of the form A^*A:

$$\omega(A^*A) \geq 0.$$

We say that linear functionals corresponding to states are positive functionals.

Proportional states are identified. It is often convenient to consider only states satisfying the condition $\omega(1) = 1$ (normalized states).

Now we can define the notion of the expectation value of a physical quantity in a given state. For a polynomial function $A \mapsto f(A) \in \mathcal{A}$ on $\mathcal{A}$ the mathematical expectation (the avarage) is given by the formula:

$$\langle f(A) \rangle_\omega = \omega(f(A)).$$

For a C^*-algebra, we can define $f(A)$ for any continuous function f of a self-adjoint element A by utilizing the fact that a continuous function on a bounded set can be represented as a uniform limit of polynomials. Consequently, knowing the expectation values for all continuous functions, we can define the concept of the probability distribution of a physical quantity in a normalized state (a state for which $\omega(1) = 1$).

The concept of a state alone is insufficient: one also requires the concept of evolution, because the objective of physics, like any science, is to make predictions. A physicist primarily considers the problem: if an initial state is known, what is the method to predict what will happen afterward?

In the algebraic approach, to define the concept of evolution one should first consider the group $Aut(\mathcal{A})$ of automorphisms of the $*$-algebra $\mathcal{A}$. Remember that automorphisms of $\mathcal{A}$ must always commute with involution, hence the group of automorphisms, naturally acting on linear functionals, transforms positive functionals into positive ones (states into states). States must depend on time. There must be an evolution operator $U(t)$ that transforms a state at the initial moment into a state $\omega(t)$ at some other time t; in other words, $\omega(t) = U(t)\omega(0)$.

In the algebraic approach, we can assume that the operators $U(t)$ come from automorphisms of the algebra $\mathcal{A}$ denoted by the same symbol $U(t)$. As in textbook quantum mechanics, there is the Schrödinger picture, where the state evolves, and there is the Heisenberg picture, where the operator evolves. These two pictures are equivalent:

$$\omega(t)(A) = \omega(A(t)).$$

(Observing the dynamics of a state ω when an algebra element A is fixed is the same as observing the dynamics of an algebra element when the state does not change.)

In physics, the evolution operator is usually calculated from the equation of motion describing the same evolution operator, but over infinitesimal time. If there is invariance with respect to the time shift, then it can be argued that the operator describing a change over an infinitesimal time interval is itself independent of time. It has already been said that a change over finite time must be an automorphism of algebra, so changes over infinitesimal time intervals are infinitesimal automorphisms. Knowing an infinitesimal automorphism H, we solve the equation of motion $dU/dt = HU$. The solution obeying $U(0) = 1$ can be written in the form $U(t) = e^{Ht} \in \mathcal{A}$. As a result, we obtain a one-parameter group of automorphisms consisting of transformations of the form e^{Ht} (evolution operators). The state $\omega(t) = U(t)\omega(0)$ obeys the equation of motion

$$\frac{d\omega}{dt} = H\omega(t).$$

The operator H is an analog of Hamiltonian in quantum mechanics; we say that H is a "Hamiltonian".

So far, we did not give a formal definition of an infinitesimal automorphism. One possibility is: an infinitesimal automorphism is a tangent vector to a curve in an automorphism group at the unit element of this group. We will require a little more, namely, that this curve is a one-parameter subgroup of $Aut(\mathcal{A})$.

It is important to note that an infinitesimal automorphism is a derivation (see Appendix A.1 for a brief review of notions we use). Indeed, let $(U(t))_{t\in\mathbb{R}}$ be a one-parameter subgroup of the automorphisms of $\mathcal{A}$ and H - its tangent vector at $U(0) = 1$ (in other words, $\frac{dU}{dt}(0) = H$). For any $x, y \in \mathcal{A}$, $U(t)(xy) = U(t)x \cdot U(t)y$, differentiating which with respect to t and setting $t = 0$ one finds $Hx \cdot y + x \cdot Hy = H(xy)$, which is the Leibniz rule.

Conversely, if A is a derivation, that is, if the Leibniz rule is satisfied, and in addition, it is consistent with involution, that is, the condition $(Ax)^* = A(x^*)$ is satisfied, then we can hope that A is an infinitesimal automorphism. To check that A is an infinitesimal automorphism, we need to solve the equation

$$dU/dt = AU,$$

where $U(t)$ is an element of the algebra $\mathcal{A}$. If this equation has a solution with the initial condition $U(0) = 1$, then A is an infinitesimal automorphism. It can play the role of a "Hamiltonian" (as in textbook quantum mechanics where any self-adjoint operator can play the role of a Hamiltonian). If the algebra is finite-dimensional, we can apply the existence theorem for solutions of differential equations. In this case, the notions of derivation and infinitesimal automorphism are equivalent. Since algebras we consider are infinite-dimensional, in our situation not every derivation defines an infinitesimal automorphism.

It is easy to check that a commutator of two derivations is a derivation, hence derivations form a Lie algebra. The same is true for derivations that agree with involution. One can say that derivations consistent with involution form the Lie algebra of the group of automorphisms $Aut(\mathcal{A})$. For the case of infinite-dimensional groups, the notion of Lie algebra is not very well defined, but it is an important notion that works in many cases.

We considered the case when the equation of motion does not depend on time but this is not necessary. The "Hamiltonian" may depend on time, and then the equation of motion for the evolution operators has the form:

$$\frac{dU}{dt} = H(t)U(t).$$

If the operator $H(t)$ does not depend on t, then the evolution operators form a one-parameter group:

$$U(t+\tau) = U(t)U(\tau).$$

In textbook quantum mechanics any density matrix K corresponds to the linear functional $A \mapsto \omega(A) = \mathrm{Tr} K A$ on the algebra of bounded operators; this functional satisfies the condition $\omega(A^*A) \geq 0$, therefore it is a state. (The positivity condition follows from the positive definiteness of the operator K). The evolution of the density matrix is described by an equation in which the right-hand side is a commutator with a self-adjoint operator (up to a constant factor). This equation has the form $dK/dt = H(K)$, where $H(K) = [\hat{H}, K]/i\hbar$, $\hat{H}$ is a Hamiltonian of textbook quantum mechanics, and H is a "Hamiltonian".

Here we introduced the following notations: operators in Hilbert space are operators with a hat, and operators acting on density matrices are operators without a hat.

According to Stone's theorem, (not necessarily bounded) self-adjoint operators in Hilbert space correspond to one-parameter subgroups of the group of unitary operators. In Stone's theorem, the subgroups are continuous in the strong sense. If a self-adjoint operator is bounded, then the corresponding one-parameter subgroup is differentiable in the sense of norm convergence.

In what follows we will not pay attention to these subtleties.

In the geometric approach the starting point is the set of states $\mathcal{N}$. We assume that one can consider mixture of states, therefore we suppose that $\mathcal{N}$ a convex closed subset of Banach space (or of topological vector space) $\mathcal{L}$. We say that an invertible linear operator is an automorphism of $\mathcal{N}$ if it maps $\mathcal{N}$ onto itself. We fix a subgroup $\mathcal{V}$ of the group of automorphisms of $\mathcal{N}$. We assume that the evolution of a state is specified by an evolution operators $U(t) \in \mathcal{V}$ obeying the equation

$$\frac{dU}{dt} = HU$$

where H is a linear operator in $\mathcal{L}$ called "Hamiltonian".

1.3 Gelfand-Naimark-Segal (GNS) Construction

Let us show that every state (considered as positive functional on $*$-algebra) can be represented by a vector in Hilbert space. The corresponding construction is attributed to Gelfand, Naimark, and Segal and is outlined below.

Why is the representation of states as vectors in Hilbert space not always convenient? The reason is that, for the same algebra of observables, it becomes necessary to consider different Hilbert spaces.

Let us say a few more words about the relation between the algebraic approach and the standard approach based on Hilbert spaces and explain why the algebraic approach is better. Suppose we have an involution-preserving representation of the algebra $\mathcal{A}$ by operators in the Hilbert space $\mathcal{H}$. In other words, we consider an involution-preserving homomorphism of the algebra $\mathcal{A}$ into an algebra of operators. Let us denote the operator corresponding to the element A of the algebra $\mathcal{A}$ by $\hat{A}$, then each normalized vector $\Phi \in \mathcal{H}$ specifies a normalized state ω of the algebra $\mathcal{A}$ by the formula

$$\omega(A) = \langle \Phi, \hat{A}\Phi \rangle. \tag{1.3}$$

(Moreover, each density matrix K specifies a state according to the formula $\omega(A) = \mathrm{Tr}(K\hat{A})$). In other words, it is possible to obtain states from vectors in Hilbert space.

Can all states be obtained this way? The answer to this question is positive. Every state can be represented by a vector in Hilbert space, and this is the reason why physicists can work all the time in Hilbert spaces. However, not every two states can be represented by vectors in the same Hilbert space. (Recall that we consider only separable Hilbert spaces.) For example, in statistical physics, we consider equilibrium states. Each equilibrium state lies in its own Hilbert space. This is not always convenient.

One Hilbert space, as a rule, is sufficient in quantum field theory, because there we usually consider a Hilbert space, in which the ground state lies. Its elements correspond to excitations of the ground state. In quantum field theory usually we consider only excitations of the ground state. However, it is impossible to use only one Hilbert space in quantum electrodynamics (see Sect. 4.9).

We are going to prove that every state of a $*$-algebra is represented by a vector from a Hilbert space. We will construct a pre-Hilbert space $\mathcal{E}$ for each algebra $\mathcal{A}$ and a state ω. (Here it is convenient to work with pre-Hilbert spaces.) We will define a representation $A \to \hat{A}$ of the algebra by operators in pre-Hilbert space $\mathcal{E}$ in such a way that some cyclic vector denoted by $\theta \in \mathcal{E}$, will correspond to the state ω. (This means that $\omega(A) = \langle \theta, \hat{A}\theta \rangle$.) The fact that a vector is cyclic means that any other vector can be obtained from it using operators from algebra (all vectors have the form $\hat{A}\theta$, where $A \in \mathcal{A}$).

Our construction is unambiguous (up to equivalence), as will be seen from the proof. Let us assume that we already have the representation we need. We can define the scalar product in $\mathcal{A}$ by the formula

$$\langle A, B \rangle = \omega(A^* B).$$

Knowing this scalar product in algebra, we can calculate the scalar product of vectors $\hat{A}\theta$ and $\hat{B}\theta$:

$$\langle \hat{A}\theta, \hat{B}\theta \rangle = \langle \theta, \hat{A}^* \hat{B}\theta \rangle = \langle \theta, \widehat{(A^* B)}\theta \rangle = \omega(A^* B).$$

Since the vector θ is cyclic, each vector of the space $\mathcal{E}$ has the form $\hat{A}\theta$. We can say that there is a mapping $\nu : \mathcal{A} \to \mathcal{E}$, which transforms A into $\hat{A}\theta$, and this mapping is surjective. It follows that $\mathcal{E}$ is obtained from $\mathcal{A}$ by factorization. We need to factorize over all vectors that give 0 in the scalar product with any other vectors (null vectors).

Now we can answer the question: how to construct $\mathcal{E}$ and the cyclic vector θ from the algebra $\mathcal{A}$ and ω? One should take the algebra $\mathcal{A}$, introduce the scalar product in it as $\omega(A^* B)$, and factorize with respect to null vectors. We obtain a pre-Hilbert space $\mathcal{E}$. (The scalar product in $\mathcal{A}$ descends to a scalar product in quotient space). The cyclic vector θ is the equivalence class of the identity element in $\mathcal{A}$.

We did two things: first, we built a pre-Hilbert space, and second, we proved that our construction is essentially unique, nothing else can be done. (This follows from cyclicity). This reasoning (the Gelfand-Naimark-Segal or GNS construction) is very important. We will use this construction many times. Instead of pre-Hilbert space, we can consider its completion-Hilbert space $\bar{\mathcal{E}}$ (then vector θ will be cyclic in a weaker sense: vectors of type $\hat{A}\theta$ will be dense in $\bar{\mathcal{E}}$).

To illustrate, let us take some stationary state (a state that does not change during evolution) and apply the GNS construction to it. Then we get some Hilbert space and a cyclic vector in it, which also will be stationary (will not depend on time).

Assertion: if we start with a stationary state, then the evolution operators $U(t)$ descend to the Hilbert space.

This is very easy to understand. In GNS construction we used $\omega(A^* B)$ as a scalar product. But this scalar product is invariant with respect to operators $U(t)$ because ω is invariant (that is, it is not changed by evolution operators). Since the scalar product is invariant, the operators $U(t)$ descend into unitary operators $\hat{U}(t)$. The operators $\hat{U}(t)$ form a one-parameter group. It has a generator (infinitesimal automorphism) $\hat{H}$, and this is what in physics is called the Hamiltonian. (Actually, this is not exactly true, because in physics the Hamiltonian is assumed to be a self-adjoint operator, -therefore we need an imaginary unit in the definition: $\hat{U}(t) = e^{-i\hat{H}t}$.)

We say that ω is a ground state if the spectrum of the operator $\hat{H}$ is non-negative. Note that the ground state will have zero energy under this definition. If we apply the GNS construction to the algebra of bounded operators and the state corresponding to the eigenvector of the Hamiltonian $\hat{H}$ with eigenvalue E, then the generator of the group $\hat{U}(t)$ constructed with the GNS construction can be identified with $\hat{H} - E$. In quantum field theory, we always count the energy as the difference from the energy of the ground state and ignore the infinite contribution to the ground state energy. The algebraic GNS construction does this automatically.

1.4 Algebraic Approach to Classical Mechanics

We have examined the algebraic approah to quantum mechanics. This method is equally effective in classical mechanics. To demonstrate this, we will utilize the Hamiltonian formalism and apply the same logic. A pure state is characterized by generalized momenta $p = (p_1, ..., p_n)$ and generalized coordinates $q = (q^1, ..., q^n)$, which represent points in a $2n$-dimensional space known as phase space. This is a pure state, but similar to quantum mechanics, mixed states can also be considered.

A mixed state is a probability distribution on phase space, i.e., a positive measure on phase space (the measure of the entire space is assumed to be equal to 1). All these probability distributions form a convex set. Pure states are the extreme points of this set. A pure state is a probability distribution that is supported at exactly one point. It is described by a probability density, which is a delta function. Any function is a superposition of delta functions (in other words, any function can be represented as an integral of delta functions). This implies that any probability distribution on phase space corresponds to a probability distribution on pure states, and pure states can be identified with extreme points of the set of all states.

In classical mechanics, every state can be represented in a unique way as a mixture of pure states. This differentiates classical mechanics from quantum mechanics, where a state can be represented as a mixture of pure states in multiple ways. It will be later explained how quantum mechanics can be derived from classical mechanics by limiting the set of observables. From a physical perspective, it is natural to assume that our devices can only measure a subset of observables. It will be demonstrated that in such a situation, classical mechanics can lead to quantum mechanics (Sect. 4.2).

The equations of motion in Hamiltonian formalism can be written as

$$\frac{dq}{dt} = \frac{\partial H}{\partial p}, \frac{dp}{dt} = -\frac{\partial H}{\partial q}.$$

The Liouville equation for the probability density function can be written in terms of Poisson brackets as:

$$\frac{d}{dt}\rho(p, q, t) = \{H, \rho(p, q, t)\}.$$

The Poisson brackets in this case are defined by the formula

$$\{f, H\} = -\frac{\partial f}{\partial p}\frac{\partial H}{\partial q} + \frac{\partial f}{\partial q}\frac{\partial H}{\partial p}.$$

To determine the evolution operator $U(t)$, it is necessary to solve the equation $\frac{d}{dt}U(t) = LU(t)$, where $L\rho = \{H, \rho\}$. This is a different form of the Liouville equation.

In classical mechanics, just as in quantum mechanics, one can alternatively track the evolution of observables (instead of states). The observables are functions $f(p, q)$

on the phase space. One can easily deduce from Hamilton equations that the evolution of observables is governed by the equation

$$\frac{d}{dt}f(p(t), q(t)) = -\frac{\partial f}{\partial p}\frac{\partial H}{\partial q} + \frac{\partial f}{\partial q}\frac{\partial H}{\partial p}.$$

In other words,

$$\frac{d}{dt}f(p(t), q(t)) = \{f, H\}.$$

We observe that quantum mechanics is very similar to classical mechanics in Hamiltonian formalism. We have observables that can be multiplied. They form an algebra $\mathcal{A}$. There is a notion of involution. (In classical mechanics the involution is simply complex conjugation). Each state ω corresponds to a linear functional on the algebra of observables $\mathcal{A}$: in classical mechanics, we take the integral of the function with respect to the probability distribution considered as a measure on phase space:

$$\omega(f) = \int f\omega.$$

This functional satisfies the positivity condition: $\omega(f) \geq 0$ if $f \geq 0$. Functions of the form A^*A are certainly positive (just the square of the modulus), hence the functional $\omega(f)$ is a state in the sense of the algebraic approach. We see that classical mechanics enters as a small piece into the algebraic approach to quantum mechanics, however in classical mechanics the algebra of observables is commutative.

1.5 Quantum Mechanics as a Deformation of Classical Mechanics. Weyl Algebra

Quantum mechanics, when the Planck constant is small, simplifies to classical mechanics. This suggests that quantum mechanics is a deformation, a slight alteration of classical mechanics. We should have a family of algebras $\mathcal{A}_\hbar$ that depends on the Planck constant $\hbar$. If the Planck constant $\hbar$ equals zero, we should obtain classical mechanics, i.e., we should get a commutative algebra with product $A \cdot B$. Let's assume that all these algebras are defined on the same vector space, i.e., addition and scalar multiplication are independent of the Planck constant, and the multiplication $A \cdot_\hbar B$ of elements of the algebra $\mathcal{A}_\hbar$ depends on the Planck constant. Now consider a commutator in this algebra as a function of the Planck constant:

$$[A, B]_\hbar = A \cdot_\hbar B - B \cdot_\hbar A$$

Our primary requirement is that this commutator tends to zero when the Planck constant tends to zero: $\hbar \to 0$. Let's assume that the dependence on the Planck constant is smooth. This implies that the commutator can be represented as an expression linear with respect to $\hbar$ plus an element of the algebra vanishing as $\hbar^2$:

$$[A, B]_\hbar = i\{A, B\}\hbar + O(\hbar^2).$$

It is straightforward to prove that the linear part has the same properties as the Poisson bracket. This means that the operation $(A, B) \mapsto \{A, B\} \in \mathcal{A}_0$ is a derivation with respect to both arguments (satisfies Leibniz rule):

$$\{A \cdot B, C\} = \{A, C\} \cdot B + A \cdot \{B, C\}$$

and, additionally, it satisfies the axioms of the Lie algebra. To prove this fact we use the following properties of the commutator in associative algebra:

$$[A, B] = -[B, A],$$

$$[A, [B, C]] + [B, [C, A]] + [C, [A, B]] = 0,$$

$$[AB, C] = [A, C]B + A[B, C].$$

These equations must be satisfied for each Planck constant $\hbar$. Let's decompose all these equations with respect to Planck constant. In the second equality, it is necessary to decompose to the second order, and in the rest—to the first order. Equating the leading terms with respect to $\hbar$, we obtain the desired properties.

We established that in the limit $\hbar \to 0$, the commutator divided by $i\hbar$ converges to an operation having the properties of the Poisson bracket. (In other words, in the limit $\hbar \to 0$ we obtain a Poisson algebra.) This means that the quantum mechanical Heisenberg equations of motion (1.2) become the classical Hamiltonian equations of motion in the limit $\hbar \to 0$. We have proven that in the limit $\hbar \to 0$ we obtain classical mechanics from quantum mechanics.

Let's see if it is possible to go the other way around. We wonder: can we get quantum mechanics from classical mechanics? To do this, let's first describe all possible Poisson brackets in the case when the algebra $\mathcal{A}$ is an algebra of polynomial functions on some vector space with coordinates $(u^1, ..., u^n)$. To calculate the Poisson bracket we need only to know the Poisson bracket of coordinates. This is because we deal with polynomials, and the Leibnitz rule allows one to calculate the Poisson bracket of products. A polynomial is a linear combination of the products of the coordinates, hence the bracket of two polynomials can be calculated. The result of the calculation is as follows:

$$\{A, B\} = \frac{1}{2}\sigma^{kl}(u)\frac{\partial A}{\partial u^k}\frac{\partial B}{\partial u^l},$$

where $\sigma^{kl}(u)$ denotes the Poisson bracket of coordinates u^k, u^l.

If σ is antisymmetric and independent of u, one can check that this expression satisfies the conditions imposed on the Poisson bracket. Representing σ as a block-diagonal matrix with two-dimensional blocks we obtain the standard Poisson bracket in the phase space.

Now we can ask the question: how to deform the Poisson bracket to get a family of associative algebras? This problem is not easy. It was solved relatively recently by Kontsevich [15]. But in the case when the Poisson bracket of two coordinates does not depend on u we can get a family of associative algebras imposing relations

$$\hat{u}^k\hat{u}^l - \hat{u}^l\hat{u}^l = i\hbar\sigma^{kl}. \tag{1.4}$$

Of course, we could impose these relations when σ depends on u, but then we do not know whether we would get an associative algebra. If σ does not depend on u, then we get an associative algebra, which is called a Weyl algebra. If we start with polynomials, this is essentially the only way to deform the Poisson bracket. Up until now, we worked with associative algebras, whereas in the algebraic approach, $*$-algebras are required. One can introduce the involution in the Weyl algebra by assuming the generators $\hat{u}^k$ to be self-adjoint.

We obtained commutation relations, which in a slightly different form are well-known from quantum mechanics textbooks. To show this, we require that the matrix σ be non-degenerate. This is an antisymmetric matrix; it can be written in a suitable basis as a block-diagonal matrix consisting of two-dimensional blocks. If we take advantage of this, we can reduce the commutation relations in Weyl algebra to the commutation relations

$$\hat{p}_k\hat{p}_l = \hat{p}_l\hat{p}_k, \quad \hat{q}^k\hat{q}^l = \hat{q}^l\hat{q}^k, \quad \hat{p}_k\hat{q}^l - \hat{q}^l\hat{p}_k = \frac{\hbar}{i}\delta_k^l.$$

These are commutation relations for coordinates and momenta in the standard exposition of quantum mechanics. They are called canonical commutation relations (CCR).

Instead of self-adjoint generators, one can take other generators that are not self-adjoint but are adjoint to each other and satisfy the relations $\hat{a}_k\hat{a}_l = \hat{a}_l\hat{a}_k$, $\hat{a}_k^*\hat{a}_l^* = \hat{a}_l^*\hat{a}_k^*$, $\hat{a}_k\hat{a}_l^* - \hat{a}_l^*\hat{a}_k = \hbar\delta_{kl}$. For example, one can take $\hat{a}_k = \frac{1}{\sqrt{2}}(\hat{q}^k + i\hat{p}_k)$, $\hat{a}_k^* = \frac{1}{\sqrt{2}}(\hat{q}^k - i\hat{p}_k)$.

This is how the creation and annihilation operators are denoted, but, so far, it is a formal mathematical object. We introduced them formally and wrote commutation relations for them. These commutation relations are also called canonical commutation relations.

Can we say that the resulting algebra $\mathcal{A}_\hbar$ is a deformation of the commutative algebra? Formally, we can't, because when we introduced the notion of deformation, we required that all these algebras be defined on the same space. (Otherwise, it would be difficult to consider all of them simultaneously.) Our commutative algebra consisted of polynomials and the elements of the new algebra are noncommutative polynomials—finite linear combinations of monomials built out of generators p and q. However, there is a basis in the algebra such that the algebra becomes isomorphic to the space of polynomials as a linear space. Such a basis can be constructed as follows.

Because of the commutation relations, we can shift all the $\hat{q}^k$ generators to the left and $\hat{p}^k$ to the right (or vice versa) and then remove the hats from them. Then we get a regular polynomial and we can say that the element from our algebra is represented by this polynomial, which is called a q-p-symbol of the element of Weyl algebra. Now the algebra is defined on the space of polynomials. This is not a very good representation because it is not consistent with involution. Nevertheless, it is very useful in many cases.

If we start with generators $\hat{a}_k, \hat{a}_k^*$, we can use the same idea: shift $\hat{a}_k^*$ to the left, $\hat{a}_k$ to the right, and you get what is called the normal form of a Weyl algebra element. Now you can remove the hats and get a polynomial, which is called the Wick symbol. Physicists do not use the term "Wick symbol," but they use the words"normal form" all the time. The Wick symbols agree with involution. (Involution in algebra corresponds to a complex conjugation of Wick symbols.)

We can consider a Weyl algebra with an infinite number of generators. So far we have considered the parameter k to be discrete (although the number of u_k or a_k with the index k could be infinite), but we can consider this parameter to be continuous. For example, consider an algebra with generators $\hat{a}(k), \hat{a}^*(l)$ and relations

$$\hat{a}(k)\hat{a}(l) = \hat{a}(l)\hat{a}(k), \quad \hat{a}^*(k)\hat{a}^*(l) = \hat{a}^*(l)\hat{a}^*(k),$$

$$\hat{a}(k)\hat{a}^*(l) - \hat{a}^*(l)\hat{a}(k) = \hbar\delta(k, l).$$

In this case, instead of the Kronecker symbol, we use its continuous counterpart: the δ−function. Since the function $\delta(k, l)$ is a generalized function, the generators $\hat{a}(k), \hat{a}^*(l)$ must also be treated as generalized functions. A generalized function is a function that only makes sense under the integral sign with a test function. Only the expressions $\hat{a}(f) = \int f(k)\hat{a}(k)dk$ and $\hat{a}^*(g) = \int g(l)\hat{a}^*(l)dl$ that represent formal integrals are well-defined elements of the algebra. These elements depend linearly on f and g, respectively, and satisfy the commutation relations: $\hat{a}(f)\hat{a}(g) = \hat{a}(g)\hat{a}(f)$, $\hat{a}^*(f)\hat{a}^*(g) = \hat{a}^*(g)\hat{a}^*(f)$, $\hat{a}(f)\hat{a}^*(g) - \hat{a}^*(g)\hat{a}(f) = \hbar\langle \bar{f}, g\rangle$. For these relations to make sense, the scalar product of test functions must be defined (hence the space of test functions should be a pre- Hilbert space). Since the scalar product depends on f antilinearly, we write $\bar{f}$ in the last formula. For simplicity, we will usually assume that the index k is discrete. As in the case of a finite number of degrees of freedom, every element A of Weyl algebra can be expressed in normal form (creation operators are moved to the left).

Transition to symbols is an operation that is closely related to the operation of quantization. What is quantization? Starting with classical Hamiltonian we want to obtain its quantum counterpart. If the Hamiltonian depends on u_k, then simply replacing u by $\hat{u}$ creates a problem, in what order to put these generators? In the classics, the order is not important: u_1u_2 and u_2u_1 are the same, but if we go to quantum mechanics, to Weyl algebra, the result depends on the order. This is what is called "ordering ambiguity" and means that there is no unambiguous quantization procedure. It can be made unambiguous by choosing the notion of a symbol, but there are many symbols.

There are such cases when the quantum Hamiltonian has a natural definition. Let us consider, for example, a standard situation in classical mechanics when the Hamiltonian is represented as a sum of kinetic and potential energies. The kinetic energy depends only on momenta, and the potential energy depends only on coordinates, "putting hats" we obtain a well-defined element of Weyl algebra because quantum momenta commute with each other and quantum coordinates commute with each other.

To write the equation of motion in the Heisenberg picture there is a standard way: in the classical equations of motion in place of Poisson brackets one must write commutators:

$$\frac{\partial \hat{u}^k}{dt} = i[\hat{H}, \hat{u}^k]. \tag{1.5}$$

In this formula, we take $\hbar = 1$. In what follows, Planck constant $\hbar$ will always be taken as equal to one, unless otherwise stated.

The equation of motion is meaningful if the Hamiltonian $\hat{H}$ is an element of a Weyl algebra. We know already that the equation of motion must contain an operation that satisfies the Leibniz rule (a derivation). The commutator of the form $D_h(a) = [h, a]$ satisfies the Leibniz rule $[h, ab] = [h, a]b + a[h, b]$ for any algebra and so as long as $\hat{H}$ is an element of a Weyl algebra, everything is fine. The only problem is that $\hat{H}$ is very often not an element of a Weyl algebra in the case of an infinite number of degrees of freedom. A typical example is the Hamiltonian of the form

$$\hat{H} = \sum \epsilon_k \hat{a}_k^* \hat{a}_k. \tag{1.6}$$

When the number of indices is infinite, it is an infinite sum. Such a Hamiltonian is a formal expression not belonging to the Weyl algebra. Nevertheless, one can formally take the commutator of the Hamiltonian $\hat{H}$ and $\hat{a}_k$. We obtain the equation of motion, which we will encounter more than once:

$$\frac{d\hat{a}_k}{dt} = -i\epsilon_k \hat{a}_k, \quad \frac{d\hat{a}_k^*}{dt} = i\epsilon_k \hat{a}_k^*.$$

Notice that the above equations make perfect sense—their right-hand sides do belong to the Weyl algebra. Therefore, the commutator of the Hamiltonian (1.6) with any element of the Weyl algebra belongs to the Weyl algebra. It defines a derivation on the Weyl algebra.

So, in the case of an infinite number of degrees of freedom, one can consider formal Hamiltonians of the form

$$\hat{H} = \sum \Gamma_{m,n}(k_1, ..., k_m, l_1, ..., l_n)\hat{a}^*_{k_1}...\hat{a}^*_{k_m}\hat{a}_{l_1}...\hat{a}_{l_n},$$

as long as they define derivations on the algebra. Let us stress again that a formal Hamiltonian may not be an element of the Weyl algebra, but it must make sense under the commutator sign in the equations of motion. This does not always happen, but there are very simple conditions when it makes sense. When a commutator of $\hat{a}_k$ or $\hat{a}^*_k$ with a product is taken, one should commute with each factor of that product. Due to the Kronecker symbol in CCR the commutator will get contribution only from coefficients where one of the indices coincides with k. If for any index there exist only a finite number of nonzero coefficients in the Hamiltonian containing this index, the equations of motion make sense.

Notice that one can modify the definition of Weyl algebra in many different ways taking a completion of the algebra we considered in some topology. The class of admissible formal Hamiltonians depends on the choice of completion.

1.6 Quadratic Hamiltonians

Let us consider quadratic Hamiltonians of the form

$$\hat{H}(u) = \frac{1}{2}H_{kl}\hat{u}^k\hat{u}^l,$$

where the operators $\hat{u}$ satisfy the commutation relations (1.4). Here the ordering plays no role because by changing the order we obtain an irrelevant constant. (The Hamiltonian is used only under the commutator sign, where the constant disappears.) Classical equations of motion and quantum equations of motion are exactly the same. Moreover, if we know how to solve the classical equations of motion, we immediately know how to solve the quantum equations of motion, because all equations of motion are linear, and the difference between classical and quantum mechanics arises only when the operators are multiplied.

The same problem can be made even simpler, namely, it is possible to diagonalize the Hamiltonian: If we assume that the matrix H_{kl} is positive definite, we can represent the Hamiltonian as a sum of squares. Further, we can simplify the matrix σ, preserving the representation of the Hamiltonian as a sum of squares. In order to preserve this property, only orthogonal transformations should be considered. Note that σ is an antisymmetric matrix. Therefore, it can be brought to a block-diagonal form where the block size is equal to two.

These two-dimensional blocks will be antisymmetric matrices, so the Hamiltonian will take the form of a sum of Hamiltonians of the form

$$\hat{H} = \frac{1}{2}(\hat{p}^2 + \epsilon^2\hat{q}^2),$$

where $[\hat{p}, \hat{q}] = \frac{1}{i}$. This is an extremely important simplification, which can always be done for a positive quadratic Hamiltonian. We discussed this in the case of a finite number of degrees of freedom. It is important to note that in the case of an infinite number of degrees of freedom, there exists a similar simplification. The only difference is that when a self-adjoint operator is diagonalized, a continuous spectrum may appear in addition to a discrete spectrum. We will come back to this.

Let us consider the Hamiltonian $\hat{H} = \sum \frac{1}{2}(\hat{p}_k^2 + \epsilon_k^2(\hat{q}^k)^2)$ assuming that $\epsilon_k > 0$ and solve the corresponding equations of motion

$$\frac{d\hat{p}_k}{dt} = -\epsilon_k^2 \hat{q}^k, \quad \frac{d\hat{q}^k}{dt} = \hat{p}_k.$$

These are the standard equations of motion of harmonic oscillator exactly as in classical mechanics. They can be solved in dozens of ways, but the simplest way is to introduce new variables

$$\hat{a}_k = \frac{1}{\sqrt{2}}(\sqrt{\epsilon_k}\hat{q}^k + \frac{i\hat{p}_k}{\sqrt{\epsilon_k}}), \quad \hat{a}_k^* = \frac{1}{\sqrt{2}}(\sqrt{\epsilon_k}\hat{q}^k - \frac{i\hat{p}_k}{\sqrt{\epsilon_k}}).$$

In these variables the equations of motion

$$\frac{d\hat{a}_k}{dt} = -i\epsilon_k \hat{a}_k, \quad \frac{d\hat{a}_k^*}{dt} = i\epsilon_k \hat{a}_k^*,$$

correspond to the Hamiltonian

$$\hat{H} = \sum \epsilon_k \hat{a}_k^* \hat{a}_k. \tag{1.7}$$

They have a very simple solution:

$$\hat{a}_k(t) = e^{-it\epsilon_k}\hat{a}_k(0), \quad \hat{a}_k^*(t) = e^{it\epsilon_k}\hat{a}_k^*(0).$$

In the case of an infinite number of degrees of freedom, by virtue of the spectral theorem, one can also assume that everything is diagonal, but instead of the sum we get an integral and the Hamiltonian will have the form:

$$\hat{H} = \int d\lambda \epsilon(\lambda)\hat{a}^*(\lambda)\hat{a}(\lambda).$$

In physics, λ usually consists of continuous and discrete indices, and the integral involves both integration and summation over a discrete index. In the special case when the theory is translation-invariant, we assume that the operators $\hat{a}^*(\mathbf{x})$ and $\hat{a}(\mathbf{x})$ depend on coordinates $\mathbf{x}$, which can be shifted without changing the Hamiltonian. This means that the Hamiltonian has the form

$$\hat{H} = \int d\mathbf{x}d\mathbf{y}\epsilon(\mathbf{x}-\mathbf{y})\hat{a}^*(\mathbf{x})\hat{a}(\mathbf{y}),$$

where the coefficient depends only on the difference $\mathbf{x} - \mathbf{y}$. (There may also be discrete indices in the expression in question—we should sum over these indices.)

It is possible to pass to the momentum representation (take the Fourier transform). Then the Hamiltonian will take a form:

$$\hat{H} = \int d\mathbf{k}\epsilon(\mathbf{k})\hat{a}^*(\mathbf{k})\hat{a}(\mathbf{k}).$$

We will consider translation-invariant Hamiltonians all the time, and this formula will be essentially used.

1.7 Stationary States

Now let us briefly discuss stationary (time-independent) states. If the evolution operators are denoted by $U(t)$, then a stationary state ω obeys $U(t)\omega = \omega$.

If we work in the formalism of density matrices K, then, using the fact that the equations of motion for the density matrix are written as a commutator with the Hamiltonian $\hat{H}$, we conclude that the density matrix represents a stationary state if it commutes with the Hamiltonian. In particular, if the density matrix is a function of the operator $\hat{H}$, then the state is stationary.

If we speak about pure state represented by a vector Ψ in Hilbert space, the stationary state satisfies the condition $\hat{H}\Psi = E\Psi$, i.e. the stationary state is an eigenvector of the Hamiltonian. The Hamiltonian can be interpreted as the energy operator, and E is the energy level. The vector Ψ changes with time but the state does not change as $\Psi(t) = \hat{U}(t)\Psi = e^{-itE}\Psi$ is proportional to Ψ.

In the algebraic approach, the Hamiltonian can be a formal expression, but it is possible to apply the GNS construction to stationary state ω and obtain a Hilbert space in which there are unitary operators describing the evolution (time shift). The generator of the time translation group has the meaning of the Hamiltonian (energy operator). Its eigenvalues can be interpreted as energy levels of excitations of the state ω. More precisely, such interpretation will be perfectly correct when ω itself is stationary and translation-invariant (invariant with respect to both spatial and time translations).

This can be illustrated as follows: a translation-invariant state can be represented as a horizontal line. Excitation must be perceived as a bump concentrated in a finite region on this horizontal line. The energy of the translation-invariant state is infinite, but the difference between the energy of the bump and the energy of the translation-invariant state can be finite.

An important and simple remark, which is explained in quantum mechanics textbooks in a less general situation, is as follows. Consider a classical Hamiltonian

$H(u)$ which has a minimum at a non-degenerate critical point. This means that the quadratic part in the Taylor expansion is positive definite; there are no zero modes. The quantum Hamiltonian is quadratic in the first approximation; in appropriate coordinates, it will have the form:

$$\hat{H} = \int d\lambda \epsilon(\lambda)\hat{a}^*(\lambda)\hat{a}(\lambda) + ...,$$

where the omitted terms are of the order three or higher with respect to $\hat{a}^*, \hat{a}$ (there can be no linear terms because we are at the critical point). The higher order terms with respect to $\hat{a}^*, \hat{a}$ are also higher order terms with respect to the Planck constant. At least in semiclassical approximation, we can neglect these terms. If we are not working in the semiclassical approximation, we can use the perturbation theory with respect to omitted terms.

1.8 Fock Space

Let us consider representations of the Weyl algebra $\mathcal{W}$ (or, what is the same, representations of canonical commutative relations). Among these representations, there is one remarkable, the simplest one, which is called the Fock representation, and the space in which it lives is called the Fock space. The Fock representation is defined in the following way: in it there exists a cyclic vector $|0\rangle$ (Fock vacuum) which is annihilated by all operators $\hat{a}_k$:

$$\hat{a}_k|0\rangle = 0.$$

(We denote by $\hat{A}$ the operator corresponding to the element $A \in \mathcal{W}$) This condition unambiguously defines the representation (up to equivalence). Sometimes the Fock vacuum will be denoted by θ.

One can obtain Fock representation using GNS construction applied to the functional $\omega(A)$ defined as a constant term in the normal form of A. This fact immediately follows from the relation $\omega(A) = \langle 0|\hat{A}|0\rangle$.

Let us describe Fock representation more explicitly. The vector $|0\rangle$ is cyclic. The definition of cyclicity depends on whether we live in pre-Hilbert or Hilbert space. If we live in pre-Hilbert space, we should get all vectors by applying algebra elements to a cyclic vector. In other words, the space is the smallest set containing the cyclic vector and invariant with respect to all operators $\hat{A}$. In Hilbert space, we should get all vector by applying algebra elements to cyclic vector and taking limits. In other words, the Hilbert space is the smallest closed set invariant with respect to operators $\hat{A}$ and containing the cyclic vector.

Consider all elements of the pre-Hilbert having the form

$$(\hat{a}^*_{k_1})^{n_1}...(\hat{a}^*_{k_s})^{n_s}|0\rangle. \tag{1.8}$$

They are obtained by applying all monomials of the operators $\hat{a}^*$ to the cyclic vector. Take all linear combinations of vectors (1.8). Can we get something new if we apply the operator $\hat{a}_k$ to these expressions? The answer is no, we will not, because it is possible to move the operator $\hat{a}_k$ to the right using commutation relations so that it acts on $|0\rangle$. Then the operator $\hat{a}_k$ will disappear due to the condition $\hat{a}_k|0\rangle = 0$. Therefore only expressions of the form (1.8) and their linear combinations belong to Fock space, if we use the definition of cyclicity appropriate for pre-Hilbert spaces.

Further, it should be noted that states (1.8) are eigenvectors of any Hamiltonian of the form $\hat{H} = \sum \epsilon_k \hat{a}_k^* \hat{a}_k$ with eigenvalues equal to $\sum n_k \epsilon_k$.

How to check this? It is easy to calculate the commutator of the Hamiltonian $\hat{H} = \sum \epsilon_k \hat{a}_k^* \hat{a}_k$ with the operator $\hat{a}_l^*$:

$$\hat{H}\hat{a}_l^* = \hat{a}_l^* \hat{H} + \epsilon_l \hat{a}_l^*.$$

To calculate the action of the Hamiltonian $\hat{H}$ on the vector (1.8) is enough, using this relation, to move the Hamiltonian to the right.

We obtained an orthogonal (but not orthonormal) basis of Fock space, which consists of eigenvectors (1.8). (To check orthogonality we use the fact that the representation is compatible with involution.)

All these formulas can be applied to the case of a multidimensional harmonic oscillator. There operators $\hat{a}_k, \hat{a}_k^*$ are called operators of creation and annihilation of quanta. In a crystal atoms somehow interact with each other, but the crystal is in a stationary state that is close to the ground state. At least in the first approximation the crystal is described by a quadratic Hamiltonian. For quanta in this situation, there is another name: phonons—quanta of sound. In the general case, we are dealing with a system of non-interacting bosons. The operators $\hat{a}_k, \hat{a}_k^*$ are called particle creation and annihilation operators and the numbers n_k in the formula for energy levels $\sum n_k \epsilon_k$ are called occupation numbers.

If we want the Fock space to be a Hilbert space, we have to take a completion.

There are advantages in both approaches. In pre-Hilbert space, the operators $\hat{a}_k, \hat{a}_k^*$ are defined everywhere. In the Hilbert space, these are unbounded operators, defined on a dense subset. However, some important states do not belong to the pre-Hilbert space.

A pre-Hilbert Fock space can be represented as a space of polynomials. Indeed, there is the following formula for the basis:

$$(\hat{a}_{k_1}^*)^{n_1} \ldots (\hat{a}_{k_s}^*)^{n_s} |0\rangle.$$

To get a polynomial, we delete$|0\rangle$ and remove the hats. We get a monomial with respect to variables a_k^*.

A linear combination of such monomials is a polynomial. That is, each element of the Fock space (which we consider to be pre-Hilbert space) can be represented by a polynomial. It is easy to calculate that the scalar product in such a representation in the form of polynomials will be given by the formula:

$$\langle F, G \rangle = \int da^* da F(a^*) G(a^*)^* e^{-a^* a}, \tag{1.9}$$

where $F(a^*)$ and $G(a^*)$ are polynomials corresponding to some vectors. Note that $G(a^*)$ stands with a star. The star applied to a twice is a again, hence $G(a^*)^*$ is a polynomial of a.

Let us check that the scalar product is written in the form (1.9). This can be calculated, but one can give simpler proof. The Fock space is uniquely defined by the existence of a cyclic vector which is annihilated by all annihilation operators provided that the creation and annihilation operators satisfy the required commutative relations. Let us consider the space of polynomials with respect to a_k^*. We define the operator $\hat{a}_k^*$ in this space as multiplication by a_k^* and the operator $\hat{a}_k$ as differentiation with respect to a_k^*. It is easy to see that multiplication and differentiation satisfy the necessary commutation relations. (If we multiply by a_k^*, then differentiate and apply the Leibniz rule, we get just what we need.) Thus we have commutation relations, there is also a cyclic vector that equals 1. It remains to check that $\hat{a}_k$ and $\hat{a}_k^*$ are adjoint to each other. (This is necessary for the involution to work correctly.) Indeed, in the formula (1.9) one can apply integration by parts to make sure that for this scalar product multiplication and differentiation are adjoint to each other. As a result, all properties of the Fock representation are satisfied, and so it is not necessary to compare the original scalar product with the new one: they necessarily coincide (up to numerical factor).

So far we have dealt with polynomials. But we still want to be able to work in Hilbert space. We should take a completion; it consists of holomorphic functions with respect to a_k^*, but only holomorphic functions having a finite norm in our scalar product belong to this space.

It is important to note that all this reasoning also applies to an infinite number of degrees of freedom. Although the integral is infinite-dimensional there, it still turns out to be well-defined.

There is another way to describe the Fock space. Polynomials are related to symmetric functions depending on discrete arguments. We can always assume that the coefficients of quadratic form are symmetric with respect to indices. For a cubic polynomial, the coefficients depend on three indices, and, again, we can impose the symmetry condition. It has to be imposed if we want to have an unambiguous representation. For polynomials of higher degree, the situation is similar. Therefore we can assume that there is a unique representation for every element of a pre-Hilbert Fock space in the form:

$$\sum_n \sum_{k_1,\ldots,k_n} f_n(k_1, \ldots, k_n) \hat{a}_{k_1}^* \ldots \hat{a}_{k_n}^* |0\rangle,$$

where $n = 0, 1, 2, \ldots$, and the coefficients f_n are symmetric with respect to indices. This means that the elements of the Fock space are represented as sequences of symmetric functions depending on an increasing number of discrete variables:

$$f_0,\ f_1(k),\ f_2(k_1,k_2),\ f_3(k_1,k_2,k_3),\ \ldots\ .$$

For the pre-Hilbert space, the elements of the Fock space are polynomials, therefore the corresponding sequences are finite. For the completion, one has to consider infinite sequences of symmetric functions f_0, $f_1(k)$, ..., which must satisfy the following condition: the norm $\sum_n n! \sum_{k_1,\ldots,k_n} |f_n(k_1,\ldots,k_n)|^2$ of an infinite sequence is finite. Usually, a sequence is written as a column (a Fock column). This construction also makes sense when k is a continuous parameter. In quantum mechanics textbooks, the Fock space is usually defined as the space consisting of columns of symmetric functions.

Equivalently we can take a Hilbert space $\mathcal{E}$ and define the Fock space as a direct sum of symmetric tensor powers of $\mathcal{E}$.

1.9 Hamiltonians Preserving the Number of Particles

In the realm of quantum theory, the operator

$$\hat{N} = \sum \hat{a}_k^* \hat{a}_k = \int d\lambda \hat{a}^*(\lambda)\hat{a}(\lambda),$$

which is either a sum over all k or an integral if the index is continuous, holds significant importance. This operator is physically interpreted as the number of particles or the number of quanta in the context of oscillators. If there exists an eigenvector X of $\hat{N}$ with an eigenvalue N, the operators a_k^*, when acting on X, increment the eigenvalue by 1, whereas a_k decreases it by 1 (i.e., they create or destroy a particle). This implies that an operator that conserves the number of particles must contain an equal number of creation and annihilation operators.

Let's now focus on quadratic Hamiltonians that possess this property.

Quadratic Hamiltonians that conserve the number of particles are crucial because they primarily govern the behavior near the minimum energy state (ground state). Consider a quadratic Hamiltonian that conserves the number of particles:

$$\hat{H} = \int dxdy A(x,y)\hat{a}^*(x)\hat{a}(y) = \langle a^*, Aa\rangle.$$

It contains products of the form a^*a, but does not include products of the form a^*a^* or aa. If the operator A has only a discrete spectrum, we can express the Hamiltonian as:

$$\hat{H} = \sum \epsilon_k \hat{a}_k^* \hat{a}_k,$$

where $\hat{a}_k = \int dx \phi_k(x)\hat{a}(x)$ corresponds to the eigenfunction $\phi_k(x)$ of the operator A.

Assume that x and y as vectors in Euclidean space and that A is expressed as: $A = -\frac{1}{2m}\Delta + \hat{U}$. (This operator appears in the Schrödinger equation, it is derived from the quantization of the classical Hamiltonian of the form $\frac{\mathbf{p}^2}{2m} + U(\mathbf{x})$.) In this case, the Hamiltonian $\hat{H}$ describes a system of non-interacting nonrelativistic identical bosons. By adding nonquadratic terms that conserve the number of particles, we obtain the Hamiltonian of the system of interacting nonrelativistic identical bosons.

It could be argued that a mathematician might have inferred quantum mechanics but didn't. The physicists, of course, did. Consider the logic involved. The mathematician understands that the observables in classical theory are simply functions on phase space. He recognizes that quantum mechanics is a somewhat distorted version of classical mechanics: everything adheres to classical laws, but in certain situations, there must be adjustments. Based on this, he proposes: let me deform the commutative algebra of functions: I will retain associativity, but introduce noncommutativity. The simplest deformation we have is a Weyl algebra. (We know that the deformation is governed by a Poisson bracket and the Weyl algebra corresponds to the Poisson bracket with constant coefficients.) The mathematician inferred that one should use Weyl algebra, and thereafter, he realizes that in the Weyl algebra one should consider the simplest Hamiltonian, which describes the behavior near the ground state. This is the quadratic Hamiltonian. The energy levels of this Hamiltonian are given by the formula:

$$\sum \epsilon_k n_k,$$

where $n_k = 0, 1, 2, \ldots$ are occupation numbers. This formula describes the energy levels of a system of non-interacting identical particles. There is no mystery in the emergence of identical particles. From the mathematician's perspective, they must appear. Non-identical particles may not necessarily exist, but identical ones always do because the simplest quadratic Hamiltonian already describes identical particles.

1.10 Representations of Weyl Algebra

We defined Fock representation of Weyl algebra. The Fock representation is irreducible (there is no other representation inside it). Are there any other the irreducible representations?

This question should be asked more precisely. In a Hilbert space operators $\hat{a}_k^*$, $\hat{a}_k$ are defined on a dense domain. One can change the domain, but define operators by the same formula. Is this the same representation? From a formal mathematical point of view, it is different but in any reasonable definition it is the same. When dealing with unbounded operators, we always have this problem. One can solve it, but it is better to work with bounded operators.

In a representation of Weyl algebra one can consider operators:

$$V_\alpha = e^{i\alpha_k \hat{u}^k},$$

where the exponent is a linear combination of self-adjoint operators $\hat{u}^k$ satisfying the commutation relations:

$$\hat{u}^k \hat{u}^l - \hat{u}^l \hat{u}^k = i\sigma^{k,l}.$$

We assume that he coefficients α_k are real, then the exponent is a self-adjoint operator multiplied by i. We obtain a unitary operator V_α. Unitary operator is bounded; it can be extended to the whole Hilbert space. It is convenient to use the operators V_α instead of unbounded operators $\hat{u}^k$. It is easy to check that

$$V_\alpha V_\beta = e^{-i\frac{1}{2}\alpha\sigma\beta} V_{\alpha+\beta}. \tag{1.10}$$

To verify (1.10) we use the formula:

$$e^X e^Y = e^{X+Y} e^{\frac{1}{2}C}, \tag{1.11}$$

where the commutator $[X, Y]$ of operators X and Y is a number, denoted here by C.

If we do not want to deal with unbounded operators, we can work with these unitary operators and with the exponential form of commutation relations (1.10). This is the exponential form of the Weyl algebra. A physicist would say that this is the same algebra, from the point of view of a mathematician it is not quite so.

There is a single irreducible representation of a Weyl algebra in the case of finite number of generators. Let us give a heuristic proof of this statement without using the exponential form of Weyl algebra.

Consider the particle number operator $\hat{N} = \sum \hat{a}_k^* \hat{a}_k$. In the case of a finite number of generators, it is a good operator. Let us take an eigenvector of this operator and start applying annihilation operators to it. (Strictly speaking we should prove that such eigenvector exists; we omit the proof.) If all annihilation operators give zero acting on the eigenvector, we will say that this is exactly the Fock vacuum we need. If there is such an operator which does not give zero, then we apply annihilation operators as many times as necessary to get such a vector $|0\rangle$ that all annihilation operators give zero acting on it: $a_k|0\rangle = 0$.

Such a vector exists because the particle number operator is positive definite and the annihilation operators decrease the number of particles. Now we take the subrepresentation containing the vector $|0\rangle$. We apply to the vector $|0\rangle$ all operators $\hat{a}_k^*$ many times. Taking linear combinations of the resulting expressions, we obtain a space that is invariant with respect to all creation and annihilation operators. A closure of this space is a subrepresentation of our representation in a Hilbert space; it will be a Fock representation because it contains a cyclic vector that is annihilated by all $\hat{a}_k$. We see that our representation contains Fock representation, but it is irreducible, hence it coincides with Fock representation.

In the case of an infinite number of generators, we will construct an example of representation that is not equivalent to Fock representation. Let us define new operators, denoted by the letter $\hat{A}_k$, in the Fock space by the formula

$$\hat{A}_k = \hat{a}_k - f_k, \quad \hat{A}_k^* = \hat{a}_k^* - \bar{f}_k$$

(for each $\hat{a}_k$ we subtract a different number). The commutation relations that we need in Weyl algebra are satisfied. Thus, we again have a representation of canonical commutation relations, a representation of Weyl algebra.

Now we will try to solve the equation $\hat{A}_k \Theta = 0$. Its solutions are eigenvectors of the operators $\hat{a}_k$. (These vectors called Poisson vectors.) They can be represented in the form

$$\Theta = e^{f\hat{a}^*}|0\rangle,$$

where $f\hat{a}^* = \sum f_k \hat{a}_k^*$ or as a function e^{fa^*} if we represent elements of Fock space as functions of a^*.

One can calculate the norm of a Poisson vector and the scalar product of two Poisson vectors (for example calculating Gaussian integral for the scalar product). The norm of Θ equals to $\exp(f^* f)$, which is finite if the sum $\sum |f_k|^2$ is finite. If the norm is finite, then the vector Θ belongs to the Fock space and the representation is equivalent to the Fock representation, but if the norm is infinite, a vector annihilated by the operators $\hat{A}_k$ does not exist hence this representation is not equivalent to the Fock representation. Moreover, it cannot contain a subrepresentation equivalent to the Fock representation.

One can generalize this construction. Let us consider operators $\hat{A}_k$, $\hat{A}_k^*$ defined by the formula

$$\hat{A}_k = \Phi_k^l \hat{a}_l + \Psi_k^l \hat{a}_l^* + f_k$$

$$\hat{A}_k^* = \bar{\Psi}_k^l \hat{a}_l + \bar{\Phi}_k^l \hat{a}_l^* + \bar{f}_k$$

where the coefficients are chosen in such a way that new operators $\hat{A}_k$, $\hat{A}_k^*$ also satisfy canonical commutation relations. The transition to $\hat{A}_k$, $\hat{A}_k^*$ is called a linear canonical transformation. The operators $\hat{A}_k$, $\hat{A}_k^*$ define a new representation of the Weyl algebra and, again, if the vector Θ which is the solution of the equation $\hat{A}_k \Theta = 0$ belongs to Fock space, then the new representation of CCR is equivalent to the Fock representation. If this condition is not satisfied, then the new representation is not equivalent to the Fock representation (see [3] for details).

Linear canonical transformations are often useful. Sometimes they and their analogs in the fermionic case are called Bogoliubov transformations.

1.11 Grassmann Algebra

A unital associative algebra with anticommuting generators is called Grassmann algebra. The elements of the Grassman algebra can be considered polynomials of anticommuting variables.

Let us consider the Grassmann algebra Λ_n with anticommuting generators $\epsilon^1, ..., \epsilon^n$:

$$\epsilon^i \epsilon^j = -\epsilon^j \epsilon^i.$$

Every element of Grassmann algebra can be represented as a sum of monomials with respect to ϵ^i:

$$\omega = \alpha + \sum_i \alpha_i \epsilon^i + \sum \alpha_{ij} \epsilon^i \epsilon^j + ... + \sum \alpha_{i_1,...,i_k} \epsilon^{i_1} ... \epsilon^{i_k} + ... \tag{1.12}$$

This representation is not unique, however, we can get a unique representation requiring antisymmetry of coefficients $\alpha_{i_1,...,i_k}$. Another standard representation of ω is based on the remark that each element of a Grassman algebra can be uniquely written as a sum of monomials in such a way that in each monomial the indices increase:

$$\omega = \alpha + \sum_i \alpha_i \epsilon^i + \sum_{i<j} \alpha_{ij} \epsilon^i \epsilon^j + ... + \sum_{i_1 < ... < i_k} \alpha_{i_1,...,i_k} \epsilon^{i_1} ... \epsilon^{i_k} + ... + \alpha_{1,...,n} \epsilon^1 ... \epsilon^n.$$

(We can use anticommutation relations to move the smaller indices to the left, and there cannot be two matching indices: if we take $i = j$ in these relations we obtain that the square of a generator is zero.)

The Grassman algebra, just like the usual algebra of polynomials, is $\mathbb{Z}$-graded: $\Lambda_n = \sum_k \Lambda_n^k$. This means that there is a notion of degree: the degree of each monomial is the number of generators in that monomial; clearly the degrees add up when monomials are multiplied, as for ordinary polynomials.

From this $\mathbb{Z}$-grading one can get $\mathbb{Z}_2$-grading by saying that there are even and odd elements: $\Lambda_n^{even} = \sum_{k \geq 0} \Lambda_n^{2k}$ and $\Lambda_n^{odd} = \sum_{k \geq 0} \Lambda_n^{2k+1}$. The $\mathbb{Z}_2$-grading is more important because it governs multiplication. An even element commutes with anything. Two odd elements anticommute.

We can say that an element of a Grassman algebra is a function of anticommuting variables; it is automatically a polynomial because there are only a finite number of monomials (in the case when we have a finite number of anticommuting variables). The analogy with functions is an important idea because it suggests that there must be an analysis in Grassmann algebra, and indeed there is. One can define differentiation $\partial_i = \partial / \partial \epsilon^i$ with respect to a variable ϵ^i. To differentiate one have to delete that variable. If there is no variable ϵ^i in the monomial, then the derivative is zero. For the case of anticommuting variables, there are notions of left derivative and right derivative. For definiteness, we will consider the left derivative; this means that before deleting the corresponding variable should be moved to the left:

$$\partial_i (\epsilon^i \epsilon^{i_1} ... \epsilon^{i_n}) = \epsilon^{i_1} ... \epsilon^{i_n},$$

$$\partial_i (\epsilon^{i_1} ... \epsilon^{i_n}) = 0$$

if $i \neq i_k$.

We calculate a derivative of a product using the Leibniz rule. In Grassmann algebra we have a modification of Leibniz rule (the graded Leibniz rule):

$$\partial_i(\omega\rho) = (\partial_i\omega)\cdot\rho + (-1)^{\bar{\omega}}\cdot\omega\cdot\partial_i\rho,$$

where $\omega, \rho \in \Lambda_n$, and ω has parity $\bar{\omega}$, that is, ω is either even ($\bar{\omega} = 0$) or odd ($\bar{\omega} = 1$). If this rule is satisfied, we speak of odd derivation; if the regular Leibniz rule is satisfied, we have even derivation.

There is also a notion of integration $\int : \Lambda_n \to \mathbb{C}$. By definition, the integral of any monomial of non-maximum degree gives zero, and the integral of a monomial of maximum degree gives plus or minus one:

$$\int \epsilon^{i_1}...\epsilon^{i_k} d^n\epsilon = 0 \quad \text{if } k < n,$$

$$\int \epsilon^1...\epsilon^n d^n\epsilon = 1.$$

In our notations we get $+1$ when the generators are ordered in ascending order. The integral of a derivative equals zero:

$$\int (\partial_i\omega) d^n\epsilon = 0.$$

(The derivative cannot contain a term of maximum degree.) From this and Leibniz rule we can derive the rule of integration by parts:

$$\int (\partial_i\omega)\cdot\rho d^n\epsilon = -(-1)^{\bar{\omega}} \int \omega\cdot\partial_i\rho d^n\epsilon.$$

When we defined a Grassman algebra Λ_n, we fixed a system of generators ϵ^1, ..., ϵ^n. Of course, it is possible to take another system of generators (this is analogous to a change of variables); then the notion of derivative will change, the notion of integral will change. For the change of variables, there is an analogue of chain rule and an analogue of the Jacobian.

Let us consider the special case where the change of variables is linear:

$$\tilde{\epsilon}^i = \sum A^i_j \epsilon^j.$$

It is easy to check that one can obtain the integral with respect to the new variables from the integral with respect to the old variables by multiplying the latter by $(\det A)^{-1}$ where A stands for the matrix A^i_j. (In conventional calculus we multiply by $\det A$.)

In a smooth function f that depends on a real variable $x \in \mathbf{R}$, we can substitute x by an even element ω of the Grassman algebra. To define $f(\omega)$ we represent ω in the form $\omega = a + \nu$, where a is a number and ν is a nilpotent element (i.e., $\nu^k = 0$ for some k). Take the Taylor series expansion of the function $f(a + \nu)$ with respect to the nilpotent part:

$$f(\omega) = f(a) + \frac{f'(a)}{1!}\nu + ... + \frac{f^l(a)}{l!}\nu^l + ...$$

and note that because of nilpotency this Taylor series has a finite number of terms.

In particular, we can consider e^ω. As an example, consider the exponent of a quadratic expression:

$$e^{\lambda_1 \epsilon^1 \epsilon^2 + \lambda_2 \epsilon^3 \epsilon^4 + ... + \lambda_k \epsilon^{2k-1} \epsilon^{2k}} = e^{\lambda_1 \epsilon^1 \epsilon^2} ... e^{\lambda_k \epsilon^{2k-1} \epsilon^{2k}} =$$

$$(1 + \lambda_1 \epsilon^1 \epsilon^2)...(1 + \lambda_k \epsilon^{2k-1} \epsilon^{2k}),$$

it follows that

$$\int e^{\lambda_1 \epsilon^1 \epsilon^2 + ... + \lambda_k \epsilon^{2k-1} \epsilon^{2k}} d^{2k}\epsilon = \lambda_1 ... \lambda_k.$$

In the more general case where $\omega = \frac{1}{2} \sum a_{ij} \epsilon^i \epsilon^j$, it is possible to represent the antisymmetric nonsingular matrix a in the block-diagonal form by changing the variables. This allows us to calculate the Gaussian integral:

$$\int e^\omega d^n \epsilon = (\det a)^{1/2}.$$

The answer is almost the same as in the usual case where we had $(\det a)^{-1/2}$ instead of $(\det a)^{1/2}$.

1.12 Clifford Algebras and Their Representation

Weyl algebra has a counterpart that is defined by the same formula, but instead of commutators we use anticommutators:

$$\hat{u}^k \hat{u}^l + \hat{u}^l \hat{u}^k = 2h^{kl}, \tag{1.13}$$

where h^{kl} is an invertible symmetric matrix. A unital associative algebra with generators obeying these relations is called Clifford algebra. The same relations appear in Dirac equation: gamma matrices are generators of a Clifford algebra.

We work with complex algebras; over complex numbers all non-degenerate quadratic forms are equivalent, hence all Clifford algebras with the same number of generators are isomorphic.

One can consider Clifford algebra as $\mathbb{Z}_2$-graded algebra (the parity of a monomial is defined as a parity of the number of generators in the monomial).

The following discussion mirrors the logic of the preceding sections, but we replace "Weyl" by "Clifford" and "commutator" by "anticommutator". Elements of Grassmann algebra play the role of complex numbers.

In what follows we consider only Clifford algebras with even number $2r$ of generators. For such algebra, we can find generators $\hat{a}_k, \hat{a}_k^*, k = 1, ..., r$ obeying equations:

$$\hat{a}_k\hat{a}_l + \hat{a}_l\hat{a}_k = 0, \quad \hat{a}_k^*\hat{a}_l^* + \hat{a}_l^*\hat{a}_k^* = 0, \quad \hat{a}_k\hat{a}_l^* + \hat{a}_l^*\hat{a}_k = \delta_{kl}, \tag{1.14}$$

which are called canonical anticommutation relations (CAR). They are obtained from the canonical commutation relations by replacing commutators with anticommutators. Note that CAR are exactly anticommutation relations that are satisfied by differentiation and multiplication with respect to generators of Grassman algebra. This remark leads to 2^r -dimensional irreducible representation of Clifford algebra (of CAR); we will prove that irreducible representation is unique (up to equivalence).

We introduce involution in the above algebra assuming that it interchanges generators $\hat{a}_k$ and $\hat{a}_k^*$.

As in the case of Weyl algebras, we can consider Clifford algebras with an infinite number of generators or with generators $\hat{a}(k), \hat{a}(k)^*$ that depend on a continuous parameter and satisfy the relations:

$$\hat{a}(k)\hat{a}(l) = -\hat{a}(l)\hat{a}(k), \quad \hat{a}^*(k)\hat{a}^*(l) = -\hat{a}^*(l)\hat{a}^*(k),$$

$$\hat{a}(k)\hat{a}^*(l) + \hat{a}^*(l)\hat{a}(k) = \delta(k, l).$$

In Clifford algebra with generators $\hat{a}_k, \hat{a}_k^*$ we have the notion of a normal form. Just as in the case of Weyl algebra, the generators $\hat{a}_k^*$ should be moved to the left, $\hat{a}_k$ goes to the right, but there is a small difference. Previously, starting with the normal form, we could define the Wick symbol just by removing the hats and treating a_k, a_k^* as complex variables. Here we define the Wick symbol by saying that by removing hats we get anticommuting variables. In other words, from an element of Clifford algebra, we get a polynomial of anticommuting variables (an element of Grassmann algebra).

Consider the formal Hamiltonian

$$\hat{H} = \sum \Gamma_{m,n}(k_1, ..., k_m, l_1, ..., l_n)\hat{a}_{k_1}^* ... \hat{a}_{k_m}^* \hat{a}_{l_1} ... \hat{a}_{l_n}.$$

Every monomial in this expression should contain an even number of generators.

When the number of variables is infinite, the Hamiltonian is usually not an element of a Clifford algebra, but commutators with generators make sense if the same conditions as for a Weyl algebra are imposed.

The definition of the Fock representation of the Clifford algebra is exactly the same as for the Weyl algebra: we require the existence of a cyclic vector $|0\rangle$ for which the condition $\hat{a}_k|0\rangle = 0$ is satisfied. Due to the cyclicity condition, one can obtain a basis by applying the operators $\hat{a}_k^*$ to the vector $|0\rangle$. All elements of pre-Hilbert Fock space will be linear combinations of monomials of the form $\Pi(\hat{a}_k^*)^{n_k}|0\rangle$.

The only difference is that the numbers n_k (occupation numbers) in these monomials can only be zero or one ($n_k = 0, 1$) because $\hat{a}_k^{*2} = 0$. This is what is called the Pauli principle (Clifford algebras play for identical fermions the same role as Weyl algebras play for identical bosons.).

As for Weyl algebras, the elements of this basis are eigenvectors of any Hamiltonian of the form

$$\sum \epsilon_k \hat{a}_k^* \hat{a}_k$$

and eigenvalues are given by exactly the same formula as in the bosonic case: $\sum n_k \epsilon_k$.

One can say that the Hamiltonian $\sum \epsilon_k \hat{a}_k^* \hat{a}_k$ describes non-interacting fermions.

Now recall that in the bosonic case to obtain a representation of elements of the Fock space by polynomials, we removed the hats in the monomials and obtained a polynomial of complex variables. Now we want to do a similar thing: remove the hats and obtain a polynomial of anticommuting variables. The form of the scalar product is exactly the same, only the integration will be over anticommuting variables. As in the case of Weyl algebra, in this representation the operator $\hat{a}_k^*$ acts as multiplication, $\hat{a}_k$ acts as differentiation; this gives the correct anticommutation relations.

The only thing left to check is that in the scalar product

$$\langle F, G\rangle = \int da^* da F(a^*) G(a^*)^* e^{-a^* a}$$

multiplication and differentiation are adjoint operators. This can be done by applying integration by parts.

If we consider polynomials alone, we get a pre-Hilbert Fock space, but we can take a completion to obtain Hilbert space. In the case of a finite number of degrees of freedom, it is not necessary to take the completion—there are only polynomials, but in the case of an infinite number of degrees of freedom, if we want to work in a Hilbert space the completion is necessary.

In pre-Hilbert Fock representation of Clifford algebra, vector can be represented as a sum of monomials with antisymmetric coefficients:

$$\sum_n \sum_{k_1,\ldots,k_n} f_n(k_1, \ldots, k_n) \hat{a}_{k_1}^* \ldots \hat{a}_{k_n}^* |0\rangle$$

while for the Weyl algebra, the coefficients are symmetric. If we are working in Hilbert space these sums can be infinite. In other words, a point of fermionic Fock space can be considered as a sequence of antisymmetric functions whereas in Fock representation of Weyl algebra it was a sequence of symmetric functions. This is the standard representation from quantum mechanics textbooks. This representation also works when k is a continuous parameter.

We can, again, consider operator

$$\hat{N} = \sum \hat{a}_k^* \hat{a}_k = \int d\lambda \hat{a}^*(\lambda)\hat{a}(\lambda)$$

(number of particles). Again $\hat{a}_k^*$ (creation operator) increases the number of particles by one , $\hat{a}_k$ (annihilation operator) decreases it.

Let us consider operators that conserve the number of particles, as in nonrelativistic quantum mechanics, starting with quadratic Hamiltonian.

$$\hat{H} = \int dxdy A(x, y)\hat{a}^*(x)\hat{a}(y) = \langle a^*, Aa\rangle.$$

In the case when A has a discrete spectrum, we can diagonalize it and write the operator $\hat{H}$ in the form

$$\hat{H} = \sum \epsilon_k \hat{a}_k^* \hat{a}_k,$$

where ϵ_k are eigenvalues, $\phi_k(x)$ are eigenfunctions of the operator A, and

$$\hat{a}_k = \int dx \phi_k(x)\hat{a}(x).$$

Taking $A = -\frac{1}{2m}\Delta + \hat{U}$, we obtained a system of non-interacting nonrelativistic bosons in the case of CCR and a systen of non-interacting nonrelativistic fermions in the case of CAR. However in general we have to assume that the operator A acts on multicomponent functions of $x \in \mathbb{R}^3$. It has the same form, but one should add summation over discrete indices.

Which canonical relations should be taken depends on how the group of rotations of three-dimensional space acts on the wave functions. The action of this group on discrete indices determines the spin of the particle. The case of half-integer spin corresponds to fermions, we must quantize using Clifford algebra, and in the case of integer spin we get bosons, hence we should use Weyl algebra. (This can be proved in the framework of relativistic theory, but not in the framework of non-relativistic quantum mechanics.) If the representation is irreducible, then spin s is determined by the number of indices r, namely $s = (r - 1)/2$. If r is odd, we should deal with Weyl algebra, if r is even, with Clifford algebra.

In order to describe interacting particles in nonrelativistic quantum mechanics, it is necessary to add terms of higher order with an equal number of creation and annihilation operators; then these terms preserve the number of particles.

In the case of a finite number of degrees of freedom, there exists a single irreducible representation of the Clifford algebra and it is isomorphic to the Fock representation.

The proof of irreducibility of the Fock representation and its uniqueness is the same as in the case of a Weyl algebra with the difference that in the case of Clifford algebra, the proof is rigorous. In the case of Weyl algebra, it was not rigorous because the operators $\hat{a}_k$ and $\hat{a}_k^*$ were unbounded operators, but in the case of Clifford algebra these operators are bounded. This follows from the relation $\hat{a}_k\hat{a}_k^* + \hat{a}_k^*\hat{a}_k = 1$, in which both summands are positive definite.

Let us study the case when the number of operators $\hat{a}_k$ and $\hat{a}_k^*$ is infinite (the case of infinite number of degrees of freedom). We can introduce new generators that still satisfy the anticommutation relations (in other words, we can consider canonical transformations). CAR are symmetric with respect to $\hat{a}_k$ and $\hat{a}_k^*$. This allows us to construct canonical transformations swapping creation and annihilation operators.

In Fock space, we take operators defined by the formula $\hat{A}_k = \hat{a}_k$ for $k \in I, \hat{A}_k = \hat{a}_k^*$ for $k \notin I$. (Here I stands for some set of indices.) The operators $\hat{A}_k, , \hat{A}_k^*$ obey CAR, hence they define a representation of the Clifford algebra. This representation is equivalent to Fock representation if we have a cyclic vector Θ which obeys $\hat{A}_k\Theta = 0$.

Let us describe the solution to the equation for Θ. We define Θ acting on $|0\rangle$ with those operators $\hat{a}_k^*$ that were changed ($k \notin I$). Acting on Θ with any operator $\hat{A}_k = \hat{a}_k^*$ with $k \notin I$, we get zero because operators $\hat{a}_k^*$ will appear twice with the same index. The operators $\hat{A}_k = \hat{a}_k$ where $k \in I$ also give zero, because they can be transferred to $|0\rangle$. Thus we have a monomial Θ which satisfies the relation $\hat{A}_k\Theta = 0$ for all k.

If we changed only a finite number of operators then Θ is a finite monomial, it belongs to the Fock space. If, however, we changed an infinite number of operators, we will get something that does not belong to the Fock space at all—a monomial of infinite degree. The new representation will not be equivalent to the Fock representation, since it does not have the cyclic vector that we need. Thus we have examples of non-equivalent representations. This construction goes back to Dirac (the famous Dirac sea).

Let us consider linear canonical transformations of the form:

$$\hat{A}_k = \Phi_k^l \hat{a}_l + \Psi_k^l \hat{a}_l^*,$$

$$\hat{A}_k^* = \bar{\Psi}_k^l \hat{a}_l + \bar{\Phi}_k^l \hat{a}_l^*.$$

Unlike the case of CCR, one cannot add numerical terms here, but everything else is the same as for the case of Weyl algebra. If we require that new operators satisfy canonical anticommutation relations, we obtain a representation of the Clifford algebra; if the equation $\hat{A}_k\Theta = 0$ has a normalized solution the new representation is equivalent to the Fock representation (see [3] for details).

Notice, in conclusion, that a transformation $\tilde{\hat{u}}^k = a_l^k \hat{u}^l$ preserving a quadratic form with coefficients h_{ab} (orthogonal transformation) specifies an automorphism of Clifford algebra; we denote it by ρ_a. From this remark, one gets instantly what is called a spinor representation of the orthogonal group $SO(2r)$. (Recall that we consider Clifford algebras with an even number of generators.) Namely, we take an irreducible representation ψ of Clifford algebra sending an element A to the operator $\psi(A)$. Then composing ψ with the automorphism ρ_a we can construct a new representation $\psi\rho_a$ that can be represented in the form $\psi\rho_a = U_a\psi U_a^{-1}$ where U_a is a unitary operator. (This follows from the uniqueness of irreducible representation.) The operators U_a are defined up to a sign. They specify a two-valued representation of the orthogonal group; this follows from the relation $\rho_{ab} = \rho_a\rho_b$. (We consider here the orthogonal group over complex numbers, but very similar considerations can be applied over real numbers.)

1.13 Adiabatic Approximation. Decoherence

Let us consider a situation when the Hamiltonian $\hat{H}(t)$ depends on time but changes adiabatically (slowly). Let us suppose that all energy levels $E_n(t)$ of the Hamiltonian $\hat{H}(t)$ where t is fixed are different (the spectrum is simple). We assume that they are differentiable with respect to t as well as the corresponding eigenvectors $\phi_n(t)$. Suppose that the time-dependent vector $\phi_n(t)$ changes slowly; this means that its derivative over t can be neglected. Let us start with the eigenvector of the Hamiltonian $\hat{H}(0)$, then we can show that during the evolution specified by slowly changing Hamiltonian $\hat{H}(t)$ it remains an eigenvector, but it will be an eigenvector of another Hamiltonian (of the Hamiltonian $\hat{H}(t)$ where t is fixed). More precisely, in the adiabatic approximation

$$\hat{U}(t)\phi_n(0) = e^{-i\alpha_n(t)}\phi_n(t), \tag{1.15}$$

where the phase factor $e^{-\alpha_n(t)}$ obeys

$$\frac{d\alpha_n(t)}{dt} = E_n(t).$$

To check this we differentiate (1.15), and neglect the derivative of $\phi_n(t)$.

We assumed that $\phi_n(t)$ changes slowly over time, this is not obvious because the eigenvector is defined only up to a constant factor. To carry out more precise consideration we suppose that the Hamiltonian $\hat{H}(g)$ depends on some parameter or many parameters, which we denote by g. Assume that the eigenvectors $\phi_n(g)$ and the eigenvalues $E_n(g)$ depend smoothly on g. Let us suppose that the parameter g depends on time in such a way that the derivative of g with respect to t can be neglected. For example, we can fix a function $g(t)$ and construct a family $g_a(t) = g(at)$. Corresponding Hamiltonians $\hat{H}_a(t) = \hat{H}(g(at))$ and their eigenvectors $\phi_n(g(at))$ vary slowly for small a. The derivative of these eigenvectors with respect to t vanishes in the limit $a \to 0$; this remark permits us to justify the above considerations.

To analyze the evolution of density matrices, we should solve the equation $\frac{dK}{dt} = H(t)K(t) = \frac{1}{i}[\hat{H}(t), K(t)]$.

The eigenvectors $\psi_{mn}(t)$ of the "Hamiltonian" $H(t)$ can be expressed in terms of the eigenvectors of the Hamiltonian $\hat{H}(t)$ denoted as $\phi_n(t)$ by the formula

$$\psi_{mn}(t)x = \langle \phi_n(t), x \rangle \phi_m(t).$$

In $\hat{H}(t)$- representation (the representation where the operator $\hat{H}(t)$ is diagonal) the vectors $\psi_{mn}(t)$ are matrices having only one non-zero entry equal to 1. Repeating the above arguments we obtain the evolution of these eigenvectors in adiabatic approximation:

$$U(t)\psi_{mn}(0) = e^{-i\beta_{mn}(t)}\psi_{mn}(t), \quad \frac{d\beta_{mn}(t)}{dt} = E_m(t) - E_n(t).$$

(We differentiate with respect to t and neglect the derivative of the vector $\psi_{mn}(t)$.) A very important fact: when $m = n$ we can assume that the phase factor is zero: $\beta_{mm} = 0$.

Let us write the density matrix K as a sum of the eigenvectors ψ_{mn} with some coefficients k_{mn}:

$$K = \sum k_{mn}\psi_{mn}.$$

Instead of considering the evolution of the eigenvectors ψ_{mn} we can consider the evolution of the coefficients $k_{mn}(t)$. The formulas are the same - coefficients $k_{mn}(t)$ get phase factors:

$$k_{mn}(t) = e^{-i\beta_{mn}(t)}k_{mn}, \quad \beta_{mn}(t) = \int_0^t (E_m(\tau) - E_n(\tau))d\tau.$$

If the adiabatic Hamiltonian $\hat{H}(t)$ is such that at time T it returns to what it was at time zero: $\hat{H}(T) = \hat{H}(0)$, then the diagonal entries of matrix K do not change, but the non-diagonal entries do—they are multiplied by a phase factor.

Let us fix now the Hamiltonian $\hat{H}$ describing some quantum system. Let's assume that the interaction with the environment changes the Hamiltonian; the new Hamiltonian $\hat{H}(t)$ may depend on time, but we assume that it changes slowly. We can imagine that a cosmic particle flies not very close to our molecule. The particle generates an electric field; this means that the Hamiltonian governing the molecule changes. If this particle flies far enough we can assume that the change is adiabatic. We do not need cosmic particles to create adiabatic electromagnetic fields, these fields are everywhere (starting with microwave cosmic radiation and ending with radiation of microwave oven in the office nearby).

We do not know these adiabatic perturbations, but we know that the diagonal elements of the density matrix are not affected by adiabatic perturbation, and the non-diagonal elements of the density matrix acquire phase factors, which we, of course, do not know, because we do not know the perturbation.

The above statements can be interpreted differently. One can consider linear combinations of the form $\alpha_0\phi_0 + \alpha_1\phi_1$ of two eigenvectors of the Hamiltonian

$\hat{H} = \hat{H}(0)$. In the evolution of this state, phase factors will appear: $\alpha_k(t) = e^{-iE_k t}\alpha_k$. These phase factors are predictable if the Hamiltonian does not depend on time, but if an adiabatic perturbation is imposed, then phase factors become unpredictable (absolute values of coefficients $\alpha_k(t)$ remain constant). Before, the two eigenvectors were coherently changing over time, but now this coherence has disappeared. This is what is called decoherence.

From these very simple considerations, one can obtain the standard prescription for the calculation of probabilities in quantum theory. Let us assume that the interaction of a molecule with the environment can be described by a random adiabatic perturbation $\hat{H}(t)$ of the Hamiltonian $\hat{H}$.

We assume that there is a Hamiltonian that depends on some parameters $\lambda \in \Lambda$ and there is some probability distribution on Λ. Let us suppose that the adiabatic perturbation acts in the period from 0 to T. Then, as we know, the entries $k_{mn}\lambda, T)$ of the density matrix $K_\lambda(T)$ in the $\hat{H}$ -representation get phase factors $C_{mn}(\lambda, T)$. The phase factors $C_{mn}(\lambda, T)$ are equal to 1 for the diagonal entries and non-trivial for other entries. Since the Hamiltonian is random, the density matrix should be averaged over the perturbation, that is, the phase factors for the non-diagonal entries should be averaged. It is quite clear that averaging these phase factors results in something that is less than 1 by absolute value. By imposing some conditions, it is easy to check that the average of the non-diagonal matrix entries will be equal to zero.

The formal proof is as follows. We included our Hamiltonian into some family of Hamiltonians $\hat{H}(g)$, where g belongs to some parameter set denoted by Λ ($g \in \Lambda$). Let us assume that all these perturbations are such that $g(0) = 0$, $g(1) = 0$ and the dependence of the Hamiltonian on time is defined by the formula $\hat{H}(g(t))$. We define the adiabatic Hamiltonian as follows:

$$\hat{H}_\alpha(t) = \hat{H}(g(\alpha t)),$$

where $\alpha \to 0$. (The time varies now from zero to $T = \alpha^{-1}$.) If we denote by $E_n(g)$ the eigenvalues of the Hamiltonian $\hat{H}(g)$, the values of phase factors $e^{-i\beta_{mn}(t)}$ at $t = T$ will be determined by the following formula:

$$\beta_{mn} = \int_0^T d\tau(E_m(g(\alpha\tau)) - E_n(g(\alpha\tau))).$$

By substituting $\alpha\tau = \tau'$ we obtain:

$$\beta_{mn} = \frac{1}{\alpha}\int_0^1 d\tau'(E_m(g(\tau')) - E_n(g(\tau')).$$

Now let us use the Riemann-Lebesgue lemma:

$$\int e^{ikx}\rho(x)dx \to 0$$

at $k \to \infty$ if $\rho(x)$ is absolutely integrable.

This lemma implies that imposing some conditions on the probability distribution on Λ we can prove that the coefficients k_{mn} of the density matrix vanish if $m \neq n$.

As a result, the density matrix K becomes diagonal in $\hat{H}$-representation due to interaction with random adiabatic perturbation. (If the initial density matrix corresponds to a pure state this effect is known as the collapse of the wave function.)

Let us now denote the diagonal matrix elements by letters p_n. We see that the averaged density matrix is a mixture of pure states with probabilities p_n. This is the usual formula for the probabilities of different pure stationary states (of eigenstates of the Hamiltonian $\hat{H}$) in a given mixed state. In particular, if we started with a pure state we obtain the standard formulas of the theory of measurements for the probabilities of different energy levels. (If a pure spate is represented as a linear combination $\sum c_n\phi_n$ of eigenvectors ϕ_n of the Hamiltonian $\hat{H}$ then the diagonal entries of the corresponding density matrix are equal to $|c_n|^2$.)

Notice that usually decoherence and the collapse of the wave function are derived from interaction with a macroscopic classical system. Here the same statements were derived from the random adiabatic interaction. Planck constant and classical systems were not used in the proof. (The proof of decoherence based on the consideration of random adiabatic interaction was given in [22]; later similar proofs were suggested in [1, 20].

Notice that using very similar arguments we can prove decoherence in the geometric approach to quantum theory. In this approach the set of states $\mathcal{N}$ is considered as a convex closed subset of Banach space $\mathcal{L}$; we assume that this set is bounded. The equation of motion has the form

$$\frac{d\sigma}{dt} = H(t)\sigma(t), \tag{1.16}$$

where $H(t)$ and $\sigma(t)$ are linear operators in $\mathcal{L}$ and the operator $\sigma(t)$ (evolution operator) maps $\mathcal{N}$ onto itself and obeys $\sigma(0) = 1$. Let us assume that the operator H in (1.16) (the "Hamiltonian") does not depend on t and can be diagonalized, i.e. there exists a basis ψ_k of $\mathcal{L}$ consisting of eigenvectors: $H\psi_k = \epsilon_k\psi_k$; it follows from boundedness of $\mathcal{N}$ that ϵ_k is purely imaginary. Let us consider adiabatic perturbation $H(t)$ of H assuming that for fixed t the operator $H(t)$ has eigenvectors $\psi_k(t)$ slowly varying with t. Denoting the corresponding eigenvalues by $\epsilon_k(t)$ and neglecting the derivative of $\psi_k(t)$ with respect to t we can check that

$$\frac{d}{dt}(\beta_k(t)\psi_k(t)) \approx H(t)(\beta_k(t)\psi_k(t))$$

if $\frac{d\beta_k(t)}{dt} = \epsilon_k(t)$. This means that $\sigma(t)\psi_k \approx \beta_k(t)\psi_k(t)$ where $\sigma(t)$ is the evolution operator obeying (1.16).

Assuming that $H(t)$ is a random adiabatic perturbation we obtain that the evolution of ψ_k is unpredictable if $\epsilon_k \neq 0$. This is the analog of decoherence in the geometric approach (see Chap. 3 for more details).

1.14 Statistical Physics

In this section, we review some notions of statistical physics.

Both in classical and quantum statistical physics one can introduce a notion of an equilibrium state and in both cases, one can define it as a state of maximum entropy under given conditions.

If a state is represented by a density matrix acting in Hilbert space $\mathcal{H}$, the entropy of the state is given by the formula

$$S = -\mathrm{Tr} K \log K.$$

If the density matrix is diagonal, then the diagonal elements p_i of the matrix are interpreted as probabilities, and this formula gives the usual expression for the entropy of probability distribution:

$$S = -\sum p_i \log p_i.$$

If we have two quantum systems with Hilbert spaces $\mathcal{H}_1$ and $\mathcal{H}_2$ then we can consider composite system with Hilbert space $\mathcal{H}_1 \otimes \mathcal{H}_2$. If a state of the first system is represented by density matrix K_1 and the state of the second system by density matrix K_2 then the state of the composite system is represented by density matrix $K_1 \otimes K_2$. The entropy of $K_1 \otimes K_2$ is equal to the sum of entropies of K_1 and K_2. (To check this one should represent K_1 and K_2 as diagonal matrices).

If the first system is governed by a Hamiltonian $\hat{H}_1$ and the second system by a Hamiltonian $\hat{H}_2$ we can assume that the composite system is governed by the Hamiltonian $\hat{H}_1 \otimes 1 + 1 \otimes \hat{H}_2$. Then we can say that we are dealing with two non-interacting systems, In this case, tensor product $\psi' \otimes \psi''$ of stationary state ψ' of $\hat{H}_1$ with eigenvalue E' and stationary state of $\hat{H}_2$ with eigenvalue E'' is a stationary state of the composite system with energy $E' + E''$.

Let us fix a Hamiltonian $\hat{H}$ acting in the space $\mathcal{H}$. We can define an equilibrium state as the state with maximum entropy for given average energy (for given expectation value of energy) $E = \mathrm{Tr} \hat{H} K$ (then we get canonical distribution; maximizing the entropy for states with energy in some interval we obtain microcanical distribution).

Maximizing the entropy for a given average energy, we obtain the density matrix

$$\frac{e^{-\beta \hat{H}}}{Z}. \tag{1.17}$$

To calculate the constant Z we notice that the density matrix, by definition, must have a trace equal to 1 and therefore this constant must be equal to the trace of the operator $e^{-\beta \hat{H}}$:

$$Z = \mathrm{Tr} e^{-\beta \hat{H}}.$$

This expression is called the statistical sum or partition function. It makes sense only if the operator $e^{-\beta\hat{H}}$ belongs to the trace class.

If the operator $\hat{H}$ has eigenvalues E_i, then $Z = \sum e^{-\beta E_i}$.

The physical meaning of β is the inverse temperature: $\beta = \frac{1}{T}$. The expression for the statistical sum shows that for $\beta \to \infty$ or, what is the same, $T \to 0$, only the term with minimum energy (the term corresponding to the ground state) contributes to this expression (we assume that the ground state is non-degenerate). We see that in the case of zero temperature the equilibrium state is a pure state (ground state).

If we have two non-interacting systems, both in equilibrium states with the same temperature, then the composite system will be in an equilibrium state with this temperature. This statement following from (1.17) supports the interpretation of β as inverse temperature.

The derivation of the formula (1.17) is based on the method of Lagrange multipliers. We assumed that the average energy $\mathrm{Tr}\hat{H}K = E$ is fixed; in addition, we know that the trace of a density matrix is equal to one: $\mathrm{Tr}K = 1$. Introducing Lagrange multipliers β and ζ we see that we should calculate stationary points of the expression

$$L = -\mathrm{Tr}K \log K - \beta\mathrm{Tr}(\hat{H}K - E) - \zeta(\mathrm{Tr}K - 1).$$

These points satisfy the equation

$$-\log K - 1 - \beta\hat{H} - \zeta = 0.$$

To verify this we must use the formula:

$$\delta\mathrm{Tr}\phi(K) = \mathrm{Tr}\phi'(K)\delta K \tag{1.18}$$

for the variation of the trace of a function of K.

The proof of this formula can be reduced to the case $\phi(K) = K^n$. In the variation of the function K^n we have n terms due to noncommutativity, but after taking the trace, all these terms become identical and we get (1.18).

Notice that the entropy is an adiabatic invariant (does not change if the evolution of the density matrix is governed by slowly changing Hamiltonian).

Although the statistical sum Z itself has no direct physical meaning, many physical quantities can be expressed in terms of Z. In particular, we can calculate the average energy

$$E = \bar{H} = -\frac{1}{\beta}\frac{\partial \log Z}{\partial \beta}$$

and entropy

$$S = \beta E + \log Z.$$

Free energy is defined by the formula $F = E - TS$; it is expressed in terms of the statistical sum as follows:

$$F = -T \log Z.$$

Instead of calculating the maximum entropy we can search for the minimum of free energy. This follows from the method of Lagrange multipliers. (The temperature T serves as a Lagrange multiplier.)

As a rule, the physical quantities can be obtained as follows: we take a family of Hamiltonians depending on some parameter and differentiate statistical sum or free energy with respect to this parameter. In particular, if the Hamiltonian $\hat{H}$ changes a little:

$$\hat{H}(\lambda) = \hat{H} + \lambda A + \dots,$$

then the new value of the statistical sum is

$$Z(\lambda) = Z + (-\beta)\lambda \mathrm{Tr} A e^{-\beta\hat{H}} + \dots.$$

The derivative of $\log Z$ with respect to λ at $\lambda = 0$ (and hence the derivative of free energy at this point) is controlled by the average value of the added term

$$\bar{A} = \frac{\mathrm{Tr} A e^{-\beta\hat{H}}}{Z}.$$

(we will also use the alternative notation $\langle A\rangle$ for this expression).

Namely,

$$\bar{A} = -T\frac{\partial \log Z}{\partial\lambda} = \frac{\partial F}{\partial\lambda} \tag{1.19}$$

(the derivatives are calculated at the point $\lambda = 0$).

Let us introduce a notion of correlation function; it will be essentially used in the next chapter. A correlation function in some state can be defined as an average of a product of some physical quantities $\langle A_1 \cdots A_n\rangle$ in this state. We can also assume that these physical quantities are time-dependent—we consider Heisenberg operators satisfying Heisenberg equations. Then the expression $\langle A_1(t_1)\cdots A_n(t_n)\rangle$ is also called a correlation function. One can consider correlation functions for any state—not necessarily an equilibrium state. If it is an equilibrium state, we write the inverse temperature value as the index:

$$\langle A_1(t_1)\dots A_n(t_n)\rangle_\beta.$$

If the Hamiltonian depends linearly on a set of parameters $\lambda_1, \dots, \lambda_k$, we can calculate the correlation functions by differentiating the free energy F. For example, if $\hat{H} = \hat{H}_0 + \lambda_1 A_1 + \dots \lambda_k A_k$ we have

$$\frac{\partial^2 F}{\partial\lambda_i\partial\lambda_j} = \langle A_i A_j\rangle - \langle A_i\rangle\langle A_j\rangle \tag{1.20}$$

(The derivatives are calculated at the point $\lambda_i = 0$.) The RHS of (1.20) is called a truncated correlation function. Higher truncated correlation functions can be defined as higher derivatives of F. They will be important later.

Note that all statements about the statistical sum and related things usually are not applicable in the case of an infinite number of degrees of freedom. In nonrelativistic quantum mechanics, usually, they are applicable when we are in a finite volume. In infinite volume they do not work—the statistical sum is not well defined, and maximum entropy is infinite. One should consider the statistical sum and correlation functions first in finite volume and after that one should take the limit of correlation functions. One cannot work directly in the infinite volume.

Note that in passing to an infinite volume one usually takes a limit not with a fixed number of particles, but with a fixed density of particles. In other words, when passing to the limit, the number of particles changes in proportion to the volume, then the density of particles remains constant.

Suppose now that the set of correlation functions in the infinite volume is obtained—what to do with it? We do not have Hilbert space in which the operators A entering the definition of correlation functions are defined in infinite volume, but there are correlation functions. Usually in such a case, one can construct a Hilbert space from these correlation functions, applying some analog of the GNS construction. Physicists usually do this implicitly. They simply say "Now we have an equilibrium state—it can be represented by a vector in a Hilbert space or a density matrix in a Hilbert space and there these operators A act". In fact, it is necessary to apply a construction which in axiomatic quantum field theory is called "reconstruction theorem". There the role of correlation functions is played by Wightman functions.

Now let us turn to the question: how can one deal in the algebraic approach with equilibrium states in a situation when it is impossible to use the maximum entropy principle? Here we can apply what is called the Kubo-Martin-Schwinger condition (KMS):

$$\langle A(t)B\rangle_\beta = \langle BA(t + i\beta)\rangle_\beta.$$

This is the condition on the correlation functions of the observables A and B in the equilibrium state. It is easy to derive in the case of finite-dimensional Hilbert space because in this case, everything is well-defined. In such a case, there is a Heisenberg operator which involves e^{iHt} in the definition, there is a density matrix in which $e^{-\beta H}$ appears. We can assume that the time t in the expression for the evolution operator is a complex number, and when it is purely imaginary with $Im(t) > 0$, then we obtain the density matrix of an equilibrium state (up to a constant factor). This is an important observation: in some sense, we obtain statistical physics from quantum dynamics in imaginary time. This is what is called "Wick rotation." In finite-dimensional case we can consider any complex time, if the dimension is infinite this is not true. However, if the correlation function $\langle BA(t)\rangle_\beta$ can be continued analytically into the strip $0 \leq Im(t) \leq \beta$, the KMS condition makes sense.

The KMS condition does not use the notion of entropy—you only need to know correlation functions. It also works in an infinite volume. One can consider the KMS condition as a definition of an equilibrium state. The KMS condition is a replacement for the maximum entropy condition in the framework of algebraic quantum theory.

Notice that the equilibrium state defined as a state obeying KMS condition is not necessarily unique. The nonuniqueness is related to phase transitions.

Now let's look at some examples.

The simplest example is the quadratic Hamiltonian.

A positive definite quadratic Hamiltonian can be reduced to the form:

$$\hat{H} = \sum \epsilon_k \hat{a}_k^* \hat{a}_k, \tag{1.21}$$

where $\hat{a}_k^*, \hat{a}_k$ obey CCR or CAR. It describes non-interacting bosons as well as multi-dimensional harmonic oscillators in the case of CCR and non-interacting fermions in the case of CAR. One can easily calculate statistical sum, average energy, etc. The statistical sum is equal to a product of statistical sums for different values of the index k. In the case of CCR for each k one should sum a geometric progression; taking the product over k we obtain

$$Z = \Pi \frac{1}{1 - e^{\beta \epsilon_k}}.$$

The average energy is calculated by the formula:

$$E = \bar{H} = \sum \epsilon_i \bar{n}_i,$$

where $\bar{n}_i = (e^{\beta \epsilon_i} - 1)^{-1}$ are the average occupation numbers.

If $\hat{a}_k^*, \hat{a}_k$ obey CAR the statistical sum is equal to

$$Z = \Pi (1 + e^{\beta \epsilon_i}),$$

and the average energy is given by the formula

$$\bar{H} = \sum \epsilon_i \bar{n}_i,$$

where $\bar{n}_i = (e^{\beta \epsilon_i} + 1)^{-1}$.

Chapter 2
Scattering

In this chapter we start with discussion of non-linear scattering, in particular, of soliton scattering in classical theories following [33–35, 37]. We continue with the definition of quantum particles and quasiparticles considered as elementary excitations of ground state and of translation-invariant stationary state. We analyze scattering matrix and its expression in terms Green's functions on shell generalizing Lehmann-Symanzic-Zimmermann (LSZ) formula and Haag-Ruelle theory [2, 9]. We define inclusive scattering matrix and show how one can express it in terms of generalized Green's functions [24, 28, 29]. **The exposition in this chapter mostly follows the review [10]. All sections, except 2.8, include reprinted material from that review that is openly available.**

2.1 Solitons as Classical Analogs of Quantum Particles

Let us discuss classical analogs of particles and quasiparticles, focusing on the notions of solitons and generalized solitons. Consider a translation-invariant Hamiltonian in an infinite-dimensional phase space M consisting of vector-valued functions $f(\mathbf{x})$, where $\mathbf{x} \in \mathbb{R}^d$ are spatial coordinates. Spatial translations act as shifts of these coordinates, and time translations are governed by a Hamiltonian that is invariant with respect to spatial translations. The corresponding equation of motion can be written as:

$$\frac{\partial f}{\partial t} = Af + B(f), \tag{2.1}$$

where A is a linear operator and B represents the nonlinear part. Assuming the nonlinear part is at least quadratic, for small f the linear part dominates. We can say that $f \equiv 0$ is a solution, and in its neighborhood, one can neglect the nonlinear part.

A. Schwarz, *Quantum Mechanics and Quantum Field Theory in Algebraic and Geometric Viewpoints*, SpringerBriefs in Physics,
https://doi.org/10.1007/978-3-031-67915-5_2

Soliton (solitary wave) is defined as a solution of the form $s(\mathbf{x} - \mathbf{v}t)$. We suppose that $s(\mathbf{x})$ tends to zero as $\mathbf{x} \to \infty$. We can visualize the solution $f \equiv 0$ as a horizontal straight line, and then the soliton is a bump moving with constant speed without changing the shape. A generalized soliton is a bump that moves, with a constant average speed, but at the same time it can pulsate, it can change its shape. We do not give a precise definition of this notion.

In Lorentz-invariant theory, by applying a Lorentz transformation to a soliton we again get a soliton. We obtain a family of solitons- solitons with different velocities. The same reasoning can be used for Galilean invariance and Galilean transformations. In both cases, we have a family of functions $s_{\mathbf{p}}(\mathbf{x} - \mathbf{a})$ that is invariant under temporal and spatial translations (here $\mathbf{p}$ denotes the momentum of soliton). This family can be considered as a symplectic manifold. A family of generalized solitons also can be considered a symplectic manifold that is invariant under temporal and spatial translations; the coordinates on this manifold are the data characterizing a (generalized) soliton.

We assume that the soliton has finite energy. (The fact that the energy is finite means, roughly speaking, that the soliton is more or less concentrated in some finite domain.)

In an old paper [37] we conjectured that for many systems and for almost all initial conditions having finite energy the solution behaves in the following way for times tending to plus or minus infinity. If there are no solitons or generalized solitons in the theory then asymptotically the solution obeys a linear equation. In the general case, we get a few solitons plus something that approximately satisfies a linear equation (a tail). This is a well-known result for integrable systems in the case $d = 1$ (see, for example, [8]); we have conjectured that this is true without the assumption of integrability in any dimension. Later this hypothesis has also been expressed in other papers. Soffer [18, 33] calls it "grand conjecture", Tao [35] calls it "soliton resolution conjecture". So far there are no results in this direction for $d > 1$ (and even for non-integrable theories in the case $d = 1$) if there exist solitons in the theory. (For the case when solitons do not exist see [34].)

This conjecture can be justified by the following reasoning. Let us assume that the initial condition is a field concentrated in some domain. In this case, we should expect the spreading of wave packet. That is, if the initial data were concentrated in some domain, then later the solution spreads to a larger domain. The energy is conserved, so this spreading causes the amplitude of the wave to decrease. If the amplitude decreases all the time, then in the case of small amplitudes the nonlinear part can be neglected, and the solution of the nonlinear equation can be approximated by a solution of a linear equation.

If there is a soliton or a generalized soliton in the theory the height of the bump remains the same, hence the amplitude does not tend to zero. However, we can expect that in the end, we get some solitons or generalized solitons plus a tail that approximately satisfies a linear equation. Of course, our reasoning is not proof, but it is convincing.

The above conjecture can be true only under some conditions. In particular, the stability of the translation-invariant state $f \equiv 0$ and of the solitons is necessary, otherwise, the solution can blow up. Nevertheless, it seems that the conjecture is true in many cases.

In these cases, there is a notion of soliton scattering. For solvable models of dimension 1+1 (one space dimension and one time dimension) this is a well-known fact. Two solitons collide, we see something that does not resemble any solitons ("a mess"), and then the same solitons appear again. The situation in the general case is slightly different: after the collision, we get some solitons (not necessarily the same solitons) plus a "tail". The tail asymptotically behaves as a solution of a linear equation.

Let us give some formal definitions. Let us denote the space of possible initial data by the letter $\mathcal{R}$. Our conjecture means that for a dense set of initial data, we can define a mapping $D^+(t) : \mathcal{R} \to \mathcal{R}_{as}$ of initial data at the moment t to asymptotic data at $t \to +\infty$. (The asymptotic data characterize the solitons and the asymptotic behavior of the tail.). We can also consider the asymptotic data at $t \to -\infty$ to get a mapping $D^-(t) : \mathcal{R} \to \mathcal{R}_{as}$.

Now we assume that there is also an inverse mapping, i.e. one can find the solution from its asymptotic behavior. That is, we want to consider inverse operators $S(t, +\infty) = (D^+(t))^{-1}$ and $S(t, -\infty) = (D^-(t))^{-1}$.

It seems that this is an interesting and not very difficult problem: to construct a solution from asymptotic data. In the quantum case, the solution to this problem is well known—it is what is called the Haag-Ruelle scattering theory; a generalization of this theory will be explained in this chapter (see [2] for the version of Haag-Ruelle theory that is close to our approach and [9] for generalization of this theory to the case when we do not have Lorentz-invariance and locality).

Now we can define the non-linear scattering matrix:

$$S = S(0, +\infty)^{-1} S(0, -\infty) : \mathcal{R}_{as} \to \mathcal{R}_{as}.$$

Roughly speaking, we fix the initial conditions at minus infinity and calculate the asymptotic behavior at plus infinity. One should expect that we can get the non-linear scattering matrix from the quantum scattering matrix in the limit $\hbar \to 0$. (More precisely, one should expect that the inclusive scattering matrix has a limit as $\hbar \to 0$ and the non-linear scattering matrix can be expressed in terms of this limit.)

Classical soliton can be considered a model of a quantum particle. In quantum field theory, the notion of a particle is an asymptotic notion: if two particles collide, we get "a mess", which then disintegrates into particles.

Notice, that the analogy with solitons makes it obvious that the existence of identical particles is not surprising.

The following considerations further emphasize the analogy of solitons with quantum particles. Consider a phase space and a Hamiltonian; in other words, we consider a symplectic manifold $\mathcal{M}$ (that can be identified with the space $\mathcal{R}$ of initial data) and an evolution operator. Assume that spatial translations act on $\mathcal{M}$ and time translations commute with spatial translations. Formally this means that on the symplectic manifold $\mathcal{M}$ we have an action of the commutative group $\mathcal{T}$ of spatial and temporal translations. Now let us take a stationary translation-invariant point $m \in \mathcal{M}$ of this symplectic manifold.

In the previous picture, such a point was the solution $f \equiv 0$.

Let us define an excitation of a translation-invariant stationary state as a state with finite energy (we assume that the energy of a translation-invariant state is equal to zero).

We define an elementary symplectic manifold $\mathcal{E}$ as such a symplectic manifold where in Darboux coordinates $\mathbf{p}, \mathbf{x}$ the spatial translations act as shifts $\mathbf{x} \to \mathbf{x} + \mathbf{a}$, while $\mathbf{p}$ does not change. We consider a Hamiltonian $\epsilon(\mathbf{p})$ that depends only on $\mathbf{p}$ (i.e. it is invariant with respect to spatial translations). Then the time translations are transformations $\mathbf{x} \to \mathbf{x} + \mathbf{v}(\mathbf{p})t$, $\mathbf{p} \to \mathbf{p}$, where $\mathbf{v}(\mathbf{p}) = \nabla\epsilon(\mathbf{p})$.

Suppose now that $\mathcal{M}$ is realized as a space of vector-valued functions $f(\mathbf{x})$ where $\mathbf{x} \in \mathbb{R}^d$ and the spatial translations act as shifts $\mathbf{x} \to \mathbf{x} + \mathbf{a}$. Let us take a symplectic embedding of the elementary symplectic space $\mathcal{E}$ into the set of excitations of translation-invariant stationary state $f(\mathbf{x}) = const$ in $\mathcal{M}$. If this embedding commutes with the space-time translations, then we get a family of solitons. To verify this we notice that symplectic embedding maps the point $(\mathbf{p}, 0)$ into some function $s_{\mathbf{p}}(\mathbf{x})$ depending on $\mathbf{p}$. Since the embedding $\mathcal{E} \to \mathcal{M}$ commutes with spatial translations, the point $(\mathbf{p}, \mathbf{a})$ maps into a shifted function $s_{\mathbf{p}}(\mathbf{x} + \mathbf{a})$. The condition that the mapping $\mathcal{E} \to \mathcal{M}$ commutes with time shifts means that the function $s_{\mathbf{p}}(\mathbf{x} - \mathbf{v}(\mathbf{p})t)$ satisfies the equation of motion.

We proved that solitons can be described in terms of a symplectic embedding of elementary symplectic space $\mathcal{E}$. In the next section we will see that quantum particles can be defined in terms of embedding of quantized elementary symplectic space.

2.2 Particles and Quasiparticles

We will introduce the notion of a particle and a more general notion of a quasiparticle (a particle is an excitation of the ground state, a quasiparticle is an excitation of any translation-invariant stationary state).

In ordinary quantum mechanics, we should have the notion of time translations T_τ. To define a notion of particle we need also a commutative group of spatial translations $T_{\mathbf{a}}$, $\mathbf{a} \in \mathbb{R}^d$ that act on states and commute with time translations. Recall that in the algebraic approach, the states are positive functionals defined up to a numerical factor, they constitute a cone denoted by C. Space-time translations must act on this cone. (In this chapter we do not use the normalization $\omega(1) = 1$.)

In the geometric approach, the space of states is the basic object, but studying scattering it is convenient to work with the cone C of non-normalized states.

We denote the commutative group of space-time translations as $\mathcal{T}$. In the algebraic approach, this group should act by automorphisms of the algebra $\mathcal{A}$. The group of automorphisms of a $*$-algebra (and hence the group $\mathcal{T}$) acts on C (we always assume that automorphisms agree with involution).

We use the standard notation, $A(\mathbf{x}, \tau) = T_{\mathbf{x}} T_{\tau} A$ for an element $A \in \mathcal{A}$ shifted in space and time. The translation-invariant stationary state ω in the algebraic approach satisfies the condition $\omega(A(\mathbf{x}, \tau)) = \omega(A)$. Standard examples of such a state are ground states and equilibrium states of translation-invariant theories.

In particular, we can consider the Weyl algebra $\mathcal{A}$ with generators $\hat{a}^*(\mathbf{x}), \hat{a}(\mathbf{x})$ obeying CCR and assume that the spatial translations simply shift the argument, while the time translations are defined by a formal Hamiltonian, which is expressed in terms of $\hat{a}^*(\mathbf{x}), \hat{a}(\mathbf{x})$ with some coefficient functions depending only on the differences $\mathbf{x}_i - \mathbf{x}_j$. This ensures translational invariance. We require that the equation of motion has a solution; this allows us to define the evolution operator. (The generators can depend also on a discrete variable; this is not reflected in our notations.)

We can do a Fourier transform and go to the momentum representation. Then the argument is denoted by $\mathbf{k}$ and a spatial translation is realized as multiplication by $\exp(i\mathbf{ka})$. The time translations will be again determined by the Hamiltonian. The condition that the functions in the coordinate representation depend on the difference leads to δ-functions corresponding to the momentum conservation, and the requirement that a coefficient function decreases rapidly as $\mathbf{x}_i - \mathbf{x}_j \to \infty$ means that the corresponding function in the momentum representation are smooth (after the δ-function is omitted).

In the geometric approach, when the group of space-time translations acts on the cone of states, we define a translation-invariant stationary state as a state that does not change under spatial and temporal shifts.

2.2.1 Excitations. Elementary Excitations

Now we want to define the notion of excitation of a translation-invariant stationary state as an analog of the previously introduced notion of a state with finite energy. When a soliton goes to infinity, we stop seeing it. Formalizing this observation we say that in the algebraic approach, a state σ is an excitation of translation-invariant state $\omega \in C$ if $(T_{\mathbf{a}}\sigma)(A)$ tends to const $\cdot\, \omega(A)$ in the limit $\mathbf{a} \to \infty$.

The constant appears here because the state is defined only up to a numerical factor.

The notion of excitation is a general notion that can be defined also in the geometric approach.

In the algebraic approach, a pre-Hilbert space $\mathcal{H}$ (it is convenient here to consider a pre-Hilbert space) can be defined for translation-invariant stationary state ω utilizing the GNS construction. The cyclic vector corresponding to the state ω will be denoted by θ. Recall that

$$\omega(A) = \langle \theta, \hat{A}\theta \rangle,$$

where $\hat{A}$ stands for the operator that corresponds to A in representation space of $*$-algebra $\mathcal{A}$.

Translations act in the algebra $\mathcal{A}$ as automorphisms. We constructed the pre-Hilbert space by factorizing the algebra in some way. This allows us to define translations in $\mathcal{H}$ as unitary operators. We define the momentum and energy operators as infinitesimal translation operators in space and time:

$$T_{\mathbf{a}} = e^{i\hat{\mathbf{P}}\mathbf{a}}, \quad T_\tau = e^{-i\hat{H}\tau}.$$

We will show that the elements of the pre-Hilbert space $\mathcal{H}$ can be considered as excitations of translation-invariant state ω. This is the explanation of the importance of GNS construction in physics.

To justify the above statements we should demand the cluster property.

Let us imagine a ferromagnetic. If spin has some direction at the origin of the coordinate system, then the same direction of spin will be more probable everywhere. This is the case when there is no correlation decay. In a more typical situation, at large distances, the spin no longer remembers the spin at the origin. We will formulate a property generalizing this observation; it is called cluster property.

Mathematically it can be expressed as follows. Let us take $\omega(A(\mathbf{x}, \tau)B)$, where A, B are two algebra elements. Then the cluster property implies that

$$\lim_{\mathbf{x}\to\infty} \omega(A(\mathbf{x}, \tau)B) = \omega(A)\omega(B)$$

This is the simplest form of cluster property. Later it will be formulated in a more general way (Sect. 2.4). At this point, we need only the following generalization. Let us take three elements B', A, and B. If A is shifted to infinity then

$$\lim_{\mathbf{x}\to\infty} \omega(B'A(\mathbf{x}, \tau)B) = \omega(A)\omega(B'B) \tag{2.2}$$

Any element of the pre-Hilbert space $\mathcal{H}$ can be represented as $\hat{B}\theta$ where $B \in \mathcal{A}$. For the state $\sigma(A)$ corresponding to the vector $\hat{B}\theta$ we have

$$\sigma(A) = \langle \hat{B}, \hat{A}\hat{B}\theta\theta \rangle = \omega(B^* AB). \tag{2.3}$$

It follows from (2.2) that

$$(T_{\mathbf{x}}\sigma)(A) = \sigma(A(0, \mathbf{x}) = \omega(B^* A(\mathbf{x}, 0)B) \to \omega(A)\omega(B^* B),$$

as $\mathbf{x} \to \infty$. This means that all elements of the pre-Hilbert space $\mathcal{H}$ correspond to excitations. In the algebraic approach, we only consider such excitations. We could start here—we could define the notion of excitation the following way: take ω, apply the GNS construction, and take the elements of the pre-Hilbert space $\mathcal{H}$.

Let us discuss the notion of elementary excitation of a translation-invariant state. Elementary excitations of the ground state are called particles. Elementary excitations of an arbitrary translation-invariant state are called quasiparticles. We will mostly use the term "elementary excitation", but we may also use the terms "particle" or "quasiparticle".

In the algebraic approach, excitations are elements of pre-Hilbert space, obtained with the GNS construction. We want to understand what should be called an elementary excitation in this situation. The notion of elementary excitation is a generalization of the notion of particle. We should be able to talk about a particle (or elementary excitation) having momentum $\mathbf{p}$. A particle could have other quantum numbers—they appear as discrete indices, which do not bother us at this moment. We will denote the vector describing a particle (or, more generally, an elementary excitation) with momentum $\mathbf{p}$ by $\Phi(\mathbf{p})$. This means that

$$\hat{\mathbf{P}}\Phi(\mathbf{p}) = \mathbf{p}\Phi(\mathbf{p}) \tag{2.4}$$

The energy of this state is some function $\epsilon(\mathbf{p})$, which is called the dispersion law:

$$\hat{H}\Phi(\mathbf{p}) = \epsilon(\mathbf{p})\Phi(\mathbf{p}). \tag{2.5}$$

Note that (2.4), (2.5) can be rewritten as

$$T_{\mathbf{a}}\Phi(\mathbf{p}) = e^{i\mathbf{pa}}\Phi(\mathbf{p}), \tag{2.6}$$

$$T_{\tau}\Phi(\mathbf{p}) = e^{-i\epsilon(\mathbf{p})\tau}\Phi(\mathbf{p}) \tag{2.7}$$

It is important to note that $\Phi(\mathbf{p})$ is not an element of Hilbert space (it has infinite norm), but a generalized vector function. To obtain an element of $\mathcal{H}$ we must consider an integral of $\Phi(\mathbf{p})$ with some test function $\phi(\mathbf{p})$:

$$\Phi(\phi) = \int d\mathbf{p}\phi(\mathbf{p})\Phi(\mathbf{p}), \tag{2.8}$$

This should be a well-defined vector. It is convenient (but not necessary) to impose normalization condition

$$\langle\Phi(\mathbf{p}), \Phi(\mathbf{p}')\rangle = \delta(\mathbf{p} - \mathbf{p}') \tag{2.9}$$

(normalization on δ-function). For vectors $\Phi(\phi)$ the normalization condition implies that

$$\langle\Phi(\phi), \Phi(\phi')\rangle = \langle\phi, \phi'\rangle. \tag{2.10}$$

We can say that an elementary excitation of the state ω is a generalized vector function $\Phi(\mathbf{p})$ taking values in the pre-Hilbert space $\mathcal{H}$ obtained by GNS construction and obeying (2.4), (2.5), (2.9).

We defined a notion of *stable* elementary excitation. Notice, however, that particles can be unstable and quasiparticles are almost always unstable. This means that our requirements are satisfied only approximately. The theory of inclusive scattering matrix developed below can be applied to unstable (quasi)particles if the lifetime of colliding (quasi)particles is much greater than the collision time. Notice that the conventional scattering matrix does not make sense for quasiparticles.

Let us define an elementary space $\mathfrak{h}$ as a subspace of the space of square-integrable functions $\phi_a(\mathbf{x})$ taking values in the space $\mathbb{C}^r$. The elements of this space can be considered as test functions; generalized functions are considered as linear functionals on $\mathfrak{h}$. In what follows we assume for definiteness that $\mathfrak{h}$ consists of smooth functions decreasing faster than any power (in other words, $\mathfrak{h}$ is the Schwartz space $\mathcal{S}$).

We will define the action of spatial and temporal translations in this space. The action of spatial translations on the test functions in the x-representation is a shift of the argument; in the momentum representation, this action is multiplication by the exponent $e^{i\mathbf{ka}}$. We can deduce the formula for time shift from the requirement that the time translations commute with spatial translations. In momentum representation the time translation T_τ is represented as multiplication by the exponent $e^{-iE(\mathbf{k})\tau}$ where $E(\mathbf{k})$ is a Hermitian matrix of dimension $r \times r$. We can diagonalize this matrix, then we get multiplication by scalar phase factors. This means that we can always restrict ourselves to the case $r = 1$.

An elementary excitation of a translation-invariant stationary state ω can be defined as commuting with spatial and temporal translations isometric mapping σ of the elementary space $\mathfrak{h}$ to the set of excitations.

It is important to note that for the scalar case $r = 1$ this definition is equivalent to the definition above. Indeed, elementary excitation has been defined as a function $\Phi(\mathbf{p})$ which is an eigenvector for momentum and energy (2.4), (2.5). There are also normalization conditions (2.9). It follows from these conditions that $\Phi(\phi)$ is a mapping of the space of test functions to the excitation space. This mapping is an isometry; this follows from the normalization condition. The formulas (2.4), (2.5) ensure that translations of the vector $\Phi(\mathbf{p})$ correspond to translations of the function $\phi(\mathbf{p})$ in both space and time. Thus, in the algebraic approach for $r = 1$ we can consider the space of test functions as elementary space and define $\sigma(\phi)$ as $\Phi(\phi)$.

It is important to emphasize that only translational invariance is important in the definition of elementary excitation, but if, say, we are dealing with a theory that is invariant with respect to rotations, it is natural to assume that the state ω is rotation-invariant. Then the group of rotations $SO(d)$ acts in the space $\mathcal{H}$. The elementary space $\mathfrak{h}$ should carry a representation of $SO(d)$ and the map of the elementary space into $\mathcal{H}$ should agree with representations of this group. (Notice that representations of $SO(d)$ can be single-valued or two-valued.) Let assume that $d = 3$ and that the generators J_1, J_2, J_3 of the Lie algebra $so(3)$ of the group $SO(3)$ can represented in the form $J_i = L_i + s_i$ where (L_1, L_2, L_3) (the operator of the orbital momentum) acts on the argument of the function $\phi_1(\mathbf{x}), ..\phi_r(\mathbf{x})$ taking values in $\mathbb{C}^r$ and (s_1, s_2, s_3) (spin operators) come from the representation of $so(3)$ in $\mathbb{C}^r$; we assume that this r-dimensional representation is irreducible. (In other words, we assume that the representation of $so(3)$ in $\mathfrak{h}$ is a tensor product of the representation in the space of scalar functions and irreducible r-dimensional representation.) In our terminology we have r particles with the same dispersion law, in standard terminology, we have a particle with a spin $s = \frac{r-1}{2}$.

In the geometric approach, we should consider cones as spaces of states. If we start with the theory in the algebraic approach, then the element ϕ of the elementary space is mapped to the state

$$(\sigma'(\phi))(A) = \langle \sigma(\phi), A\sigma(\phi)\rangle.$$

Here σ' is a quadratic (or rather, hermitian) mapping of the elementary space to the cone of states; it commutes with all translations.

This remark suggests that in the geometric approach, one should define the elementary excitation as a mapping of the elementary space to the cone of states; this mapping should commute with spatial and temporal translations.

If we are working in the algebraic approach, $\sigma(\phi)$ belongs to the pre-Hilbert space $\mathcal{H}$ and θ is a cyclic vector in this space, then $\sigma(\phi)$ is obtained by applying some element from algebra $\mathcal{A}$ to the cyclic vector:

$$\sigma(\phi) = B(\phi)\theta. \tag{2.11}$$

Then one can easily verify the formula

$$\sigma'(\phi) = L(\phi)\omega, \tag{2.12}$$

where

$$L(\phi) = \tilde{B}(\phi)B(\phi). \tag{2.13}$$

Recall that in the algebraic approach, an element $B \in \mathcal{A}$ specifies two operators on the space of functionals: one corresponds to the multiplication of the argument by B^* from the left (it is denoted $\tilde{B}$), the other corresponds to multiplication of the argument from the right (it is denoted by the same letter B).

It is convenient to include the operator $L(\phi)$ satisfying the relation (2.12) in the definition of elementary excitation in the geometric approach.

It is necessary to emphasize that the operator $B(\phi)$ exists, but it is not unique, it must be chosen somehow. We impose some conditions on it, which will allow us to develop the scattering theory. In particular, we assume that it is linear in ϕ.

The mapping $\sigma(\phi)$ considered in the algebraic situation was linear, but the mapping $\sigma'(\phi)$ is not linear at all. Indeed, from the formula (2.13) it follows that in the algebraic approach, $L(\phi)$ is a quadratic expression, or more precisely, a hermitian expression because it is linear in one variable and anti-linear in the other. (An expression f is called hermitian if it can be represented in the form $f(x) = F(x^*, x)$, where $F, y, x)$ is antilinear in the first argument and linear in the second one.) It is natural to require that in the geometric approach, $L(\phi)$ satisfies the same conditions. In what follows we use the word "quadratic" instead of "hermitian," but this is not quite precise.

If one prefers working with linear mappings, this can be done using the following general algebraic construction. For each linear complex space E in the tensor product $E \otimes \bar{E}$ of this space by a complex conjugate one can construct a cone $C(E)$ as a minimal cone containing all elements of the form $e \otimes \bar{e}$. (The bar stands for complex conjugation). We call the cone $C(\mathfrak{h})$ an elementary cone. It corresponds to the elementary space $\mathfrak{h}$, and σ' can be viewed as a linear mapping $\sigma' : C(\mathfrak{h}) \to C$ of the elementary cone to the cone of states.

Let us discuss the notion of a multi-particle state.

To simplify notations, we consider the case when the elementary space consists of scalar functions ($r = 1$).

Suppose that the support supp(ϕ) of the function $\phi(\mathbf{p})$ in the momentum space is a compact set. In that case, it is possible to find a bounded set U_ϕ for which all points of the form $\nabla\epsilon(\mathbf{p})$ where $\mathbf{p}$ belongs to supp(ϕ), are interior points (the function $\epsilon(\mathbf{p})$ is assumed smooth).

Then for large $|\tau|$ and $\frac{\mathbf{x}}{\tau} \notin U_\phi$ the function

$$(T_\tau\phi)(\mathbf{x}) = \int d\mathbf{k} e^{i\mathbf{k}\mathbf{x} - i\epsilon(\mathbf{k})\tau}\phi(\mathbf{k})$$

obeys

$$|(T_\tau\phi)(\mathbf{x})| < C_n(1 + |\mathbf{x}|^2 + \tau^2)^{-n},$$

where n is some integer (see (A.6) in Appendix A.3).

In other words, for large $|\tau|$ the function $(T_\tau\phi)(\mathbf{x})$ is small outside the set τU_ϕ, which we call the essential support of the function $(T_\tau\phi)(\mathbf{x})$.

Let us now return to the general case where the elementary space $\mathfrak{h}$ consists of vector-valued functions. We say that the set τU_ϕ is an essential support of the function

$$(T_\tau\phi)(\mathbf{x}) = \int d\mathbf{k} e^{i\mathbf{k}\mathbf{x} - iE(\mathbf{k})\tau}\phi(\mathbf{k})$$

if

$$||(T_\tau\phi)(\mathbf{x})|| < C_n(1 + |\mathbf{x}|^2 + \tau^2)^{-n},$$

at large $|\tau|$ and $\frac{\mathbf{x}}{\tau} \notin U_\phi$.

We say that functions ϕ and ϕ' do not overlap if the distance between the sets U_ϕ and $U_{\phi'}$ is positive; then the corresponding essential supports do not overlap, moreover, at large τ they are distant from each other. We say that $\phi_1, ..., \phi_n$ is a non-overlapping family of functions if ϕ_i does not overlap with ϕ_j when $i \neq j$. *We will always assume that there are many non-overlapping families of functions (more precisely linear combinations of non-overlapping families of functions should be dense everywhere in the space of families of functions we are interested in).* When $r = 1$ this is satisfied, for example, when the function $\epsilon(\mathbf{p})$ is strictly convex.

What should be called a two-particle state in the algebraic approach? When defining a one-particle space, we needed spatial and temporal shifts, but now we need more. Before, we used the representation (2.11), to describe a one-particle state with wave function ϕ. When there are two particles, B must be applied twice: $B(\phi)B(\phi')\theta$. When ϕ and ϕ' have supports far apart in coordinate space, one can say that this vector describes a state of two distant particles.*One must require that $B(\phi)$ and $B(\phi')$ almost commute with each other (then the particles will be bosonic) or almost anticommute (then the particles will be fermionic).* This definition is given in terms of states described by vectors, but it is possible to define two-particle states in terms of positive functionals on the algebra $\mathcal{A}$. For this purpose, we note that state, which corresponds to the vector $B(\phi)B(\phi')\theta$, can be written in the form $L(\phi)L(\phi')\omega$, where

$$L(\phi) = \tilde{B}(\phi)B(\phi),\ L(\phi') = \tilde{B}(\phi')B(\phi').$$

Both for bosonic and fermionic cases $L(\phi)$ almost commutes with $L(\phi')$.

In the geometric approach, a two-particle state is written as $L(\phi)L(\phi')\omega$, where $L(\phi)$ almost commutes with $L(\phi')$.

Thus, the distinction between bosons and fermions is smoothed out in the geometric approach.

In what follows we talk about bosons all the time, but the transition to fermions is trivial: one has to replace commutators with anticommutators.

2.2.2 *Particles in Non-relativistic Quantum Mechanics*

Let us consider in more detail a translation-invariant Hamiltonian of nonrelativistic quantum mechanics in Fock space. (The creation operators $\hat{a}_k^*(\mathbf{p})$ and annihilation operators $\hat{a}_k(\mathbf{p})$ where $\mathbf{p} \in \mathbb{R}^3$, $1 \leq k \leq r$ obey CCR or CAR, we work in the space of Fock representation of CCR or CAR). We can consider also creation and annihilation operators $\hat{a}^*(f), \hat{a}(g)$ where $f, g \in \mathcal{S}(\mathbb{R}^3) \otimes \mathbb{C}^r$. In this interpretation it is easy to define the action of spatial translations and rotations on $\mathcal{S}(\mathbb{R}^3) \otimes \mathbb{C}^r$ and hence on the Fock space. (In coordinate representation spatial translations are generated by the translations of $\mathbb{R}^3$. To define the action of rotations we consider the tensor product of the natural action of the group $SO(3)$ on $\mathcal{S}(\mathbb{R}^3)$ and r-dimensional representation of $SO(3)$. We assume that the r-dimensional representation is irreducible.)

If we take the operator $\hat{a}_k^*(\mathbf{p})$ and apply it to the translation-invariant Fock vacuum $|0\rangle$, we obtain an elementary excitation of the Fock vacuum:

$$\Phi_k(\mathbf{p}) = \hat{a}_k^*(\mathbf{p})|0\rangle.$$

We can say that we constructed r particles but it is better to say that we constructed a particle having spin $s = \frac{r-1}{2}$.

Besides this particle, other particles also satisfy our definition. They are called composite particles or bound states.

What is a bound state? The Hamiltonian acts on states with any number of particles and preserves this number. Take n particles and separate the motion of the center of inertia. Then the Hamiltonian can have normalizable eigenstates. They are called bound states.

Equivalently we can try to solve the Eqs. (2.4), (2.5) for $\mathbf{p} = 0$. Then the solution will contain the δ-function of the sum of momenta:

$$\int d\mathbf{p}_1 ... d\mathbf{p}_n \Psi(\mathbf{p}_1, ... \mathbf{p}_n)\delta(\mathbf{p}_1 + ... + \mathbf{p}_n)\hat{a}^*(\mathbf{p}_1)...\hat{a}^*(\mathbf{p}_n)|0\rangle.$$

If the function $\Psi(\mathbf{p}_1, ... \mathbf{p}_n)$ is square-integrable we obtain a bound state. It is easy to understand that the generalized function

$$\Phi(\mathbf{p}) = \int d\mathbf{p}_1 ... d\mathbf{p}_n \Psi(\mathbf{p}_1, ... \mathbf{p}_n)\delta(\mathbf{p} - \mathbf{p}_1 - ... - \mathbf{p}_n)\hat{a}^*(\mathbf{p}_1)...\hat{a}^*(\mathbf{p}_n)|0\rangle,$$

can be regarded as an elementary excitation. It follows from the invariance of the Hamiltonian under Galilean transformations that the energy of elementary excitation is given by the usual formula: $\epsilon(\mathbf{p}) = \mathbf{p}^2/2M + const$.

From our perspective, bound states (composite particles) are no worse than elementary excitations with $n = 1$. The general theory presented below, gives, in particular, a description of the scattering of composite particles. One can prove that non-relativistic quantum mechanics has interpretation in terms of particles in the sense of Sect. 2.3.3 (see, for example, [12]).

2.3 Scattering; *in*- and *out*-states

2.3.1 *in*- and *out*-states

Let us consider the scattering of elementary excitations in algebraic and geometric approaches. In the algebraic approach, we assume that the mapping σ of the elementary space $\mathfrak{h}$ to the space $\mathcal{H}$ defining a particle or quasiparticle can be written in the form

$$\sigma(f) = B(f)\theta. \tag{2.14}$$

In the geometric approach, we assume the existence of a mapping $L : \mathfrak{h} \to End(\mathcal{L})$, where

$$\sigma'(f) = L(f)\omega. \tag{2.15}$$

Both σ and σ' should commute with spatial and temporal translations.

Geometric approach is more general than the algebraic one; starting with the data of algebraic approach we can construct $L(f)$ by the formula $L(f) = \tilde{B}(f)B(f)$.

In both approaches, we can define states describing the scattering process. We introduce new operators for this purpose.

In the algebraic approach, we define the operator $B(f, \tau)$ by the formula:

$$B(f, \tau) = T_\tau B(T_{-\tau} f))T_{-\tau} \tag{2.16}$$

(We must remember that the time-shift acts on an operator as a conjugation with the operator T_τ.)

In the geometric approach, we define the operator $L(f, \tau)$ by similar formula:

$$L(f, \tau) = T_\tau L(T_{-\tau} f) T_{-\tau}. \tag{2.17}$$

It is easy to check that $B(f, \tau)\theta$ does not depend on τ. To verify this we notice that

$$B(f, \tau)\theta = T_\tau B(T_{-\tau} f)\theta = T_\tau \sigma(T_{-\tau} f) = \sigma(f).$$

(We used the invariance of θ with respect to time translations, the formula (2.15), and the fact that σ commutes with time translations.)

By the same reasoning, we can show that $L(f, \tau)\omega$ does not depend on τ.

We obtain that

$$\dot{B}(f, \tau)\omega = 0, \tag{2.18}$$

$$\dot{L}(f, \tau)\theta = 0, \tag{2.19}$$

where the dot at the top indicates the derivative with respect to τ. This will be our main tool.

In the case of many particles, by analogy with the definition of a single-particle state, we apply $B(f_i, \tau)$ many times with different f_i to θ:

$$\Psi(f_1, \ldots, f_n|\tau) = B(f_1, \tau)...B(f_n, \tau)\theta \tag{2.20}$$

to get a multi-particle state. Then we take the limit of the resulting expression at $\tau \to -\infty$:

$$\Psi(f_1, \ldots, f_n| -\infty) = \lim_{\tau \to -\infty} \Psi(f_1, \ldots, f_n|\tau) \tag{2.21}$$

This limit (which lies in the Hilbert space $\bar{\mathcal{H}}$, the completion of the space $\mathcal{H}$) will be called *in*-state. The physical meaning of *in*-state will be clarified below.

In the geometric approach, instead of $B(f, \tau)$ we take $L(f, \tau)$:

$$\Lambda(f_1, ..., f_n|\tau) = L(f_1, \tau), ...L(f_n, \tau)\omega \tag{2.22}$$

$$\Lambda(f_1, \ldots, f_n| -\infty) = \lim_{\tau \to -\infty} \Lambda(f_1, \ldots, f_n|\tau) \tag{2.23}$$

We obtain an *in*-state lying in $\mathcal{L}$. It corresponds to the vector $\Psi(f_1, \ldots, f_n| -\infty)$ if $L(f) = \tilde{B}(f)B(f)$.

Later in this section, we work in the algebraic approach, but the considerations in the geometric approach are very similar (see the end of this section and Sect. 4.6).

Applying the operator T_τ to $B(f, \tau')$ leads to a time shift in both argumrnts:

$$T_\tau(B(f, \tau')) = T_{\tau+\tau'}B(T_{-\tau'}f)T_{-\tau-\tau'} = B(T_\tau f, \tau + \tau'). \tag{2.24}$$

This is a purely formal calculation.

The formula (2.24) implies that

$$T_\tau \Psi(f_1, \ldots, f_n| - \infty) = \Psi(T_\tau f_1, \ldots, T_\tau f_n| - \infty). \tag{2.25}$$

If functions f_1, ..., f_n do not overlap, hence essential supports of functions $T_\tau f_i$ are far away in the limit $\tau \to -\infty$, it follows from (2.25) that in this limit $\tau \to -\infty$ the evolution of the *in*-state Ψ describes the process of scattering.

Usually one considers the scattering of particles with definite momenta. It is inconvenient to work with definite momenta in our approach because in such cases the wave functions are non-normalizable. We can consider the situation when the momentum lies in some narrow range, i.e. the support of the wave function is a small domain in the momentum space. We say that the state $T_\tau\Psi(f_1, \ldots, f_n| - \infty)$ describes a collision of particles with wave functions $(f_1, \ldots, f_n)$ if these functions do not overlap. In this case we assume that corresponding operators $B(f_i, \tau)$ almost commute for $\tau \to -\infty$, i.e. their commutator in this limit vanishes:

$$\lim_{\tau\to-\infty} ||[B(f_i, \tau), B(f_j, \tau)]|| = 0. \tag{2.26}$$

Why do we assume that? When f_i and f_j do not overlap then in the limit $\tau \to -\infty$, then the essential supports of the functions $T_\tau f_i$ and $T_\tau f_j$ are far away (see Appendix A.3). In this case, from the viewpoint of physics, it is natural to think that the corresponding operators almost commute (or almost anticommute).

The condition (2.26) can be derived from the requirement that the commutators of the two operators B that depend on the functions $\phi_a(\mathbf{x})$ and $\psi_a(\mathbf{x})$ satisfy the inequality

$$||[B(\phi), B(\psi)]|| \leq \int d\mathbf{x}d\mathbf{x}' D^{ab}(\mathbf{x} - \mathbf{x}')|\phi_a(\mathbf{x})| \cdot |\psi_b(\mathbf{x}')| \tag{2.27}$$

where $D^{ab}(\mathbf{x})$ tends to zero faster than any power when $\mathbf{x} \to \infty$. In this section we consider (2.27) as an axiom but in the next section we derive this condition from more physical requirements. Using (2.27) and assuming that the sets U_{f_i} for each pair of functions do not overlap, we obtain that the commutators in the formula (2.26) are small for large τ. This means that we can permute the operators B in the formula (2.21) for *in*-state. It follows that *in*-states are symmetric (they do not change when the arguments f_i are rearranged). Notice that we can consider anticommutators instead of commutators in (2.27); then *in*-states are antisymmetric.

Let us prove that the limit (2.21) exists. To give the proof, we additionally impose the condition of the smallness of the commutator $[\dot{B}(f_i, \tau), B(f_j, \tau)]$ at $\tau \to -\infty$, where the functions $f_i, f_j, i \neq j$ do not overlap. This is again an axiom. More precisely, we can impose the condition

$$||[\dot{B}(f_i,\tau), B(f_j,\tau)]|| \leq c(\tau), \tag{2.28}$$

where $f_i, f_j, i \neq j$ do not overlap and $c(\tau)$ is summable: $\int |c(\tau)|d\tau < \infty$. We can assume, for example, that $c(\tau) \sim 1/\tau^a$, where $a > 1$.

To prove that the *in*-state does exist (the expression $\Psi(\tau) = \Psi(f_1, \ldots, f_n|\tau)$ has a limit at $\tau \to -\infty$) we verify that $\dot{\Psi}(\tau)$ is summable. If $\dot{\Psi}(\tau)$ is summable the expression

$$\Psi(\tau_2) - \Psi(\tau_1) = \int_{\tau_1}^{\tau_2} \dot{\Psi}(\tau)d\tau. \tag{2.29}$$

tends to zero as $\tau_1, \tau_2 \to -\infty$, hence $\Psi(\tau)$ has a limit. This follows from the completeness of the space $\bar{\mathcal{H}}$ in which this vector lies.

Now we should prove that $\dot{\Psi}(\tau)$ is small. Recall that in the definition of $\Psi(\tau)$ we repeatedly applied the operator B to the state ω. Let us differentiate this expression by τ applying the Leibniz rule. We get n summands, each of which contains a derivative of one of the factors B. Now we will move $\dot{B}$ to the right, eventually moving it to the rightmost place. When we get to the very end, we use the equality $\dot{B}\theta = 0$. As a result, $\dot{\Psi}(\tau)$ will be a summable function of τ, since the commutators $[\dot{B}(f_i,\tau), B(f_j,\tau)]$ are summable.

Thus we derived the existence of limits from (2.28). This is a very important statement. It proves that we can consider the scattering of particles in our picture. We required very little, but our axioms were sufficient to prove the existence of a limit. We use this fact to introduce the notion of scattering.

It is convenient to use a slightly more general construction of the vector $\Psi(f_1, ..., f_n| -\infty)$. One can argue that the vector

$$\Psi(f_1, \tau_1, ..., f_n, \tau_n) = B(f_1, \tau_1)...B(f_n, \tau_n)\theta \tag{2.30}$$

has a limit in $\bar{\mathcal{H}}$, denoted

$$\Psi(f_1, ..., f_n| -\infty)$$

as $\tau_j \to -\infty$. (Previously we proved this statement in the case when all times τ_j are equal.)

The proof can be based on the assumption

$$||[\dot{B}(\phi), B(\psi)]|| \leq \int d\mathbf{x}d\mathbf{x}' D^{ab}(\mathbf{x}-\mathbf{x}')|\phi_a(\mathbf{x})| \cdot |\psi_b(\mathbf{x}')|,$$

where $D^{ab}(\mathbf{x}) \to 0$ is faster than any degree when $\mathbf{x} \to \infty$ (this condition is similar to the condition (2.26)).

Analogous statements can be proved if in this assumption or in (2.31) commutators are replaced by anticommutators.

The conditions we imposed on B can be obtained as a consequence of more physical requirements (for example, from the asymptotic commutativity of the algebra $\mathcal{A}$); this will be shown in Sect. 2.6.

All the above reasoning can be applied in the geometric approach. In this approach, we can impose the condition

$$||[\dot{L}(f_i, \tau), L(f_j, \tau)]|| \leq c(\tau), \tag{2.31}$$

where $c(\tau)$ is a summable function. The vector

$$\Lambda(\tau) = L(f_1, \tau)...L(f_n, \tau)\omega$$

has a limit in the Banach space $\mathcal{L}$ as $\tau \to -\infty$; the proof is the same.

2.3.2 Møller Matrices

Let us define the notion of an asymptotic bosonic Fock space $\mathcal{H}_{as}$, assuming that operators B at large distances commute. We will define an asymptotic bosonic Fock space $\mathcal{H}_{as}$ as a Fock representation of the canonical commutative relations:

$$[b(\rho), b(\rho')] = [b^*(\rho), b^*(\rho')] = 0, \quad [b(\rho), b^*(\rho')] = \langle \bar{\rho}, \rho' \rangle,$$

where $\rho, \rho' \in \mathfrak{h}$.

In the case when we consider anticommutators instead of commutators, the bosonic Fock space must be replaced by a fermionic Fock space (by the space of Fock representation of canonical anticommutation relations).

The action of spatial and temporal translations on the elementary space $\mathfrak{h}$ can be extended to the Fock space. This is clear because the n-particle part of the asymptotic space is n-th symmetric or antisymmetric power of $\mathfrak{h}$. Corresponding infinitesimal automorphisms (asymptotic Hamiltonian and asymptotic momentum operator) are quadratic with respect to creation and annihilation operators; they coincide with Hamiltonian and momentum operator on $\mathfrak{h}$ considered as the one-particle subspace of Fock space. The joint spectrum of asymptotic Hamiltonian and momentum operator coincides with the spectrum of non-interacting bosons or fermions.

Let us define the Møller matrix (half of the scattering matrix). The Møller matrix $S_\pm$ transforms a vector $b^*(f_1)...b^*(f_n)|0\rangle$ from bosonic or fermionic Fock space into the state $\Psi(f_1, ..., f_n| \pm \infty)$. It is important that the in-state is symmetric or antisymmetric. The fact that one can rearrange f_i is essential because otherwise this definition would make no sense since an n-particle subspace in a Fock space is a symmetric or antisymmetric tensor power of the space $\mathfrak{h}$. The Møller matrices are defined on a dense subset of Fock space. In the next section, it will be proved that it follows from cluster property that S_- and S_+ are isometric embeddings of $\mathcal{H}_{as}$ into $\bar{H}$, hence they can be extended to the Fock space considered as a Hilbert space. If both Møller matrices are not only isometric but also unitary, that is, they are surjective mappings of the Fock space to the entire $\bar{H}$ then we say that the theory has an interpretation in terms of particles. This means that almost every state is an *in*-state (linear combinations of *in*-states are dense everywhere). In other words, almost every state in the limit $\tau \to -\infty$ evolves to a set of distant particles; a similar statement is true for $\tau \to +\infty$.

We obtain a picture similar to the so-called soliton resolution conjecture in classical theory (Sect. 2.1).

Møller matrices commute with translations. This follows from the formula (2.25), which implies that the action of the time shift of the *in*-state corresponds to the time shift of arguments. The time shift of arguments corresponds to the time shift in Fock space, hence the formula (2.25) says that the time shift in Fock space corresponds to the time shift in Hilbert space $\bar{\mathcal{H}}$. The fact that Møller matrices commute with spatial translations is even easier to prove.

2.3.3 Scattering Matrix. *in*- and *out*-oparators

We say that the theory has particle interpretation if Møller matrices are unitary. In this case Møller matrices specify unitary equivalence between the Hamiltonian and momentum operator in $\bar{\mathcal{H}}$ and corresponding operators in the asymptotic Fock space. The same is true for the Møller matrix S_+.

The scattering matrix (S-matrix) can be defined by the formula

$$S = S_+^{-1} S_-.$$

We used a similar formula in the soliton picture (Sect. 2.1). The scattering matrix is the main object in quantum field theory.

Now we can define *in*-operators a^*_{in} using the limit of operators $B(f, \tau)$ at $\tau \to -\infty$:

$$a^*_{in}(f) = \lim_{\tau \to -\infty} B(f, \tau). \tag{2.32}$$

To prove that this is a legitimate definition we use the formula (2.30), where the operators $B(f_i, \tau_i)$ stand with different times. This means that for one of the arguments we can go to the limit later than for other arguments. Let us stress that *in*-operator is not always defined, but if in the formula (2.32) all functions $f_1, ..., f_n$ do not overlap, the *in*-operator $a^*_{in}(f_1)$ is defined on the vector $\Psi(f_2, ..., f_n| - \infty)$ and maps it into the vector $\Psi(f_1, ..., f_n| - \infty)$.

In our definition, *in*-operators depend linearly on the functions f. These operators can be regarded as generalized functions; we introduce the following notation:

$$a^*_{in}(f) = \int d\mathbf{p} f^k(\mathbf{p}) a^*_{in,k}(\mathbf{p}),$$

where $a^*_{in,k}(\mathbf{p})$ is a generalized function, and the index k specifies the particle type.

We define *out*-operators in the same way, but τ must tend to plus infinity:

$$a^*_{out}(f) = \lim_{\tau\to+\infty} B(f,\tau).$$

Notice, that *in*-operators are related to operators in asymptotic space by formulas:

$$a^*_{in}(\rho)S_- = S_- b^*(\rho), \quad S_-|0\rangle = \theta.$$

These formulas provide an alternative definition of *in*-operators. In the same way one can define operators a_{in} and a_{out} associated with annihilation operators in Fock space:

$$a_{in}(\rho)S_- = S_- b(\rho), \quad a_{out}(\rho)S_+ = S_+ b(\rho).$$

There is an obvious relation between definitions in the geometric and algebraic approaches. If the geometric approach is considered within the algebraic approach, then the operator $L(f,\tau)$ in the space of states corresponds to the operator $B(f,\tau)$ in $\bar{\mathcal{H}}$ according to the formula $L(f,\tau) = \tilde{B}(f,\tau)B(f,\tau)$.

The state $\Lambda(f_1,\ldots,f_n|\tau)$ corresponds to the vector $\Psi(f_1,\ldots,f_n|\tau)$, and $\Lambda(f_1,\ldots,f_n|-\infty)$ (*in*-state) state corresponds to $\Psi(f_1,\ldots,f_n|-\infty)$.

The analog of the Møller matrix in the geometric approach is denoted as $\tilde{S}_-$. While the Møller matrix S_- is a linear operator, $\tilde{S}_-$ is a non-linear operator. For theories that can be formulated algebraically, S_- maps a symmetric tensor power $\mathfrak{h}$ treated as a subspace of Fock space into $\bar{\mathcal{H}}$. Taking the composition of this mapping with the natural mapping $\bar{\mathcal{H}}$ to the cone of states C we obtain $\tilde{S}_-$.

When L is quadratic or hermitian, it induces a multilinear mapping of the symmetric power of the cone $C(\mathfrak{h})$ corresponding to $\mathfrak{h}$ into the cone C.

2.3.4 Inclusive Scattering Matrix

The scattering matrix describes collisions of particles. The connection of the scattering cross-section with the scattering matrix is explained in general courses in quantum mechanics and quantum field theory. We consider the connection with the inclusive cross-section; it is simpler.

Let us introduce the notion of inclusive cross-section. The scattering cross-section is related to the transition probability of, say, a pair of particles to n particles $(M,N) \to (Q_1 \ldots, Q_n)$. We will consider the process when at the end we obtain particles $(Q_1 \ldots, Q_n)$ plus something else:

$$(M,N) \to (Q_1, \ldots, Q_m, R_1, \ldots, R_n)$$

The inclusive cross-section is defined as the probability of such a process. One can obtain it from the usual cross-section summing (more precisely, integrating) over $R_1, ..., R_n$. This is true in a theory having interpretation in terms of particles, but we can define the inclusive cross-section even if there is no such interpretation: we can consider a process $(M, N) \to (Q_1, ..., Q_n + something)$, even if we do not know what is $something$.

In the geometric approach, only the inclusive cross-section makes sense. In the algebraic approach, one can work with the usual cross-sections (if the theory has interpretation in terms of particles) but it is possible (and sometimes easier) to work with the inclusive cross-sections.

Let us consider an arbitrary state ν and write the following formula for the probability density:

$$\nu(a^*_{out,k_1}(\mathbf{p}_1)a_{out,k_1}(\mathbf{p}_1)\dots a^*_{out,k_m}(\mathbf{p}_m)a_{out,k_m}(\mathbf{p}_m))$$

The expressions $a^*_{out,k_i}(\mathbf{p}_i)a_{out,k_i}(\mathbf{p}_i)$ are the numbers of particles with momentum $\mathbf{p}_i$, so the formula (2.3.4) represents the probability density in the momentum space of finding m outgoing particles of types $k_1, \dots, k_m$ with momenta $\mathbf{p}_1, \dots, \mathbf{p}_m$. We do not look at other particles.

So far ν has been any state, but now we take ν as an in-state:

$$\nu = \Lambda(g_1, ..., g_n| - \infty) = \lim_{\tau \to -\infty} L(g_1, \tau)...L(g_n, \tau)\omega$$

The in-state is determined by the incoming particles. When we defined the inclusive cross-section, we collided two particles, but it is possible to collide several particles. If in an in-state we measure the number of outgoing particles as in the formula (2.3.4), we get the inclusive cross- section by definition.

We will now represent the answer in a different form. Consider the following expression:

$$\langle 1|L(g'_1, \tau')...L(g'_{n'}, \tau')L(g_1, \tau)...L(g_n, \tau)|\omega\rangle \tag{2.33}$$

assuming that g'_i and g_j do not overlap, and the times tend to infinity, with $\tau' \to +\infty$, $\tau \to -\infty$. Acting by operators L on the state ω we obtain a linear functional on the algebra: we can calculate its value on the unit element of the algebra. Note that we used so-called *bra-ket notations*, where from the left and the right are elements from dual spaces. So, consider the expression (2.33) and denote by Q its limit as $\tau' \to +\infty$, $\tau \to -\infty$. Taking the limit $\tau \to -\infty$, we obtain

$$Q = \lim_{\tau' \to +\infty} \langle 1|L(g'_1, \tau')...L(g'_{n'}, \tau')\nu\rangle, \tag{2.34}$$

where $\nu = \Lambda(g_1, ..., g_n| - \infty)$. Since we assumed that the functions do not overlap, all commutators tend to zero and Q does not change when $g'_1, ..., g'_{n'}$ are rearranged.

Now let's look at these formulas in the algebraic approach. Then $L(g,\tau) = \tilde{B}(g,\tau)B(g,\tau)$. We have the formula $(\tilde{M}N\nu)(\alpha) = \nu(M^*\alpha N)$. (The operator $\tilde{M}$ multiplies the argument by M^* from the left, the operator N multiplies the argument from the right.) In the formula (2.34) these operators are applied to ν. They simply change the argument of ν. In addition, $\langle 1|\sigma\rangle = \sigma(1)$. The result is the expression

$$Q = \lim_{\tau'\to+\infty} \nu(B^*(g'_{n'},\tau')...B^*(g'_1,\tau')B(g'_1,\tau')...B(g'_{n'},\tau')).$$

In the limit $\tau' \to +\infty$ operators B tend to out-operators :

$$\lim_{\tau'\to+\infty} B(g,\tau') = a^*_{out}(g), \qquad \lim_{\tau'\to+\infty} B^*(g,\tau') = a_{out}(g),$$

in the limit $\tau \to -\infty$ we obtain in-operators.

Using this fact and the remark that all operators commute in the limit, we obtain the following expression for Q:

$$Q = \nu(a_{out}(g'_{n'})...a_{out}(g'_1)a^*_{out}(g'_1)...a^*_{out}(g'_{n'}))).$$

The expression

$$Q = Q(g'_1, ..., g'_{n'}, g_1, ..., g_n)$$

will be called an inclusive scattering matrix.[1] This expression is quadratic with respect to its arguments. One can switch from quadratic expressions to bilinear expressions—then the number of arguments will double. The resulting expression will also be called the inclusive scattering matrix. One can get an inclusive cross-section from it. This is not a quite trivial process. The problem is that in the definition of the inclusive scattering matrix, we considered it as a functional on non-overlapping families of functions. This functional is linear or antilinear, so it can be regarded as a generalized function, but the arguments of the generalized function (momenta) must be different. In the expression for the inclusive cross-section the momenta can coincide, hence we should take some limits in matrix elements of inclusive scattering matrix to obtain the inclusive cross-section.

In the geometric approach, we can define an inclusive scattering matrix by taking along with $\omega \in \mathcal{L}$ some translation-invariant element $\alpha \in \mathcal{L}^\vee$ of the dual space $\mathcal{L}^\vee$:

$$\lim_{\tau'\to+\infty,\tau\to-\infty} \langle\alpha|L(g_1,\tau')...L(g_m,\tau')L(f_1,\tau)...L(f_n,\tau)|\omega\rangle. \qquad (2.35)$$

The formula (2.35) can also be applied in an algebraic situation.

Notice that in this formula operators $L(g_i,\tau')$ play the role of detectors of outgoing particles.

[1] This notion was reinvented in [5] under the name "asymptotic observables". The relation to generalized Green's functions described in Sect. 2.6 also was rediscovered in [5].

It is important to notice that we should expect to have an interpretation in terms of particles only for elementary excitations of the ground state, the conventional scattering matrix makes sense only in this case. However, the inclusive scattering matrix and inclusive cross-section make sense also for almost stable quasiparticles (almost stable excitations of any translation-invariant stationary state).

2.3.5 Generalization

In the algebraic approach, we defined the notions of Møller matrices and scattering matrices using operators B obeying (2.14). In the present section, we show that in these definitions one can use B belonging to a much broader class of operators. One can prove similar statements for operators L used in the geometric approach.

To define this broader class of operators (admissible operators) we consider an element $B \in \mathcal{A}$ and the corresponding operator $B(f, \tau)$ defined by the formula (2.16).

We assume that the function $\dot{B}(f, \tau)\theta$ is summable:

$$\int d\tau ||\dot{B}(f, \tau)\theta|| < \infty \tag{2.36}$$

and that $B(f, \tau)\theta$ tends to one-particle state as $\tau \to \pm\infty$:

$$\lim_{\tau \to \pm\infty} B(f, \tau)\theta = \Phi(g). \tag{2.37}$$

where $g = f\phi$ and ϕ is a non-vanishing function. We say that an element $B \in \mathcal{A}$ obeying (2.36) and (2.37) is admissible.

Again we define a multi-particle state by the formula (2.20) and *in*- and *out*-states as limits of (2.20) as $t \to \mp\infty$. The same considerations allow us to prove the existence of these limits for admissible operator B and non-overlapping family $f_1, ..., f_n$ if the conditions (2.28) are satisfied. (Again we move the operator with a dot to the right but instead of (2.18) we use (2.36).) This statement allows us to define Møller matrices.

For example, Møller matrix $S_{\pm}$ can be defined an operator transforming a vector $b^*(g_1)...b^*(g_n)|0\rangle$ from bosonic or fermionic Fock space into the state $\Psi(f_1, ..., f_n| \pm \infty)$. Here $g_i = f_i\phi$ where ϕ is the function appearing in (2.37).

It is important to notice that on a dense subset of $S_-(\mathcal{H}_{as})$ we have

$$a^*_{out}(g) = \lim_{\tau \to +\infty} B(f, \tau), \tag{2.38}$$

where $g = f\phi$ is defined by (2.37).

Let us show that in a theory having particle interpretation almost all elements are admissible.

In such a theory any vector in $\bar{\mathcal{H}}$ has the following decomposition:

$$\sum_{r\geq 0}\int d\mathbf{p}_1...d\mathbf{p}_r c_r(\mathbf{p}_1, ..., \mathbf{p}_r)a^*_{in}(\mathbf{p}_1)...a^*_{in}(\mathbf{p}_r)\theta. \tag{2.39}$$

(This follows from similar decomposition in asymptotic space.)

Let us represent the vector $|\hat{B}|\theta\rangle$ in the form (2.39). For simplicity, assume that $\langle\theta|\hat{B}|\theta\rangle = 0$. Then we can represent $\hat{B}(f, \tau)\theta$ as

$$\hat{B}(f, \tau)\theta = \int d\mathbf{p}\phi(\mathbf{p})f(\mathbf{p})a^*_{in}(\mathbf{p})\theta +$$

$$\sum_{r\geq 2}\int d\mathbf{p}_1...d\mathbf{p}_r e^{-i\tau(\varepsilon(\mathbf{p}_1)+...\varepsilon(\mathbf{p}_r)-\varepsilon(\mathbf{p}_1+...\mathbf{p}_r))}c_r(\mathbf{p}_1, ..., \mathbf{p}_r)f(\mathbf{p}_1 + ... + \mathbf{p}_r)a^*_{in}(\mathbf{p}_1)...a^*_{in}(\mathbf{p}_r)\theta.$$

In this formula, the summation should be over all r, but for $r = 1$ due to cancellation in the exponent the time dependence disappears. The contribution of $r = 1$ gives the first summand. In most cases the sum over $r \geq 2$ tends to zero because the exponent $-i\tau(\varepsilon(\mathbf{p}_1) + ...\varepsilon(\mathbf{p}_r) - \varepsilon(\mathbf{p}_1 + ...\mathbf{p}_r))$ tends to infinity as $\tau \to \infty$ for almost all values of the arguments. (If we assume that the energy and momentum conservation laws forbid the decay of the particle, as it is assumed in Sect. 2.6.1, the exponent always tends to infinity. If B is a smooth operator as defined in Sect. 2.6.2 the coefficient functions c_r are smooth.) We obtain the relation (2.37) with $g(p) = \phi(\mathbf{p})f(\mathbf{p})$ together with the fact that the projection of $\hat{B}(f, \tau)\theta$ on one-particle space is equal to $\Phi(g)$.

Differentiating the expression for $\hat{B}(f, \tau)\theta$ with respect to τ we get (2.36) using similar reasoning.

This means that in theories having particle interpretation almost all elements $B \in \mathcal{A}$ are admissible. (It seems that all operators encountered in physics are admissible.)

Let us prove that Møller matrices (hence the scattering matrix) do not depend on the choice of an admissible operator B. Notice first of all that our proof of the existence of limit can be applied to the expression

$$\lim_{t\to\pm\infty} B_1(f_1, t)...B_n(f_n, t)\theta$$

where B_i are admissible operators obeying the condition (2.28). We define Møller matrices using this expression. Using (2.37) we obtain that the rightmost operator B_n can be replaced with any other admissible operator without changing the Møller matrix. On the other side using (2.28) we can move any operator B_i to the right; this allows us to replace it with an arbitrary admissible operator without changing the Møller matrix.

The same logic can be applied to prove the existence of the limit of the expression

$$\lim_{t_i\to\pm\infty} B_1(f_1, t_1)...B_n(f_n, t_n)\theta$$

where all times tend to infinity independently.

2.4 Correlation Functions. Cluster Property

Let us define correlation functions of some elements of algebra $\mathcal{A}$ in a translaion-invatiant state ω. (We assume that spatial and temporal translations act on $\mathcal{A}$ transforming $A \in \mathcal{A}$ into $A(\mathbf{x}, t)$.)

Take some elements $A_1 \ldots A_r \in \mathcal{A}$, and shift them in space and time. By multiplying them, we get an element of the algebra and after that, we apply ω or, what is the same, we take the average (the expectation value) of this product in the state ω. We say that

$$w_n(\mathbf{x}_1, t_1, \ldots \mathbf{x}_n, t_n) = \omega(A_1(\mathbf{x}_1, t_1) \cdots A_n(\mathbf{x}_n, t_n)) = \langle A_1(\mathbf{x}_1, t_1) \cdots A_n(\mathbf{x}_n, t_n)\rangle$$

is a correlation function. (In Sect. 1.14 correlation functions were defined in simpler situations when we have only temporal translations.)

Let us define the notion of a truncated correlation function

$$w_n^T(\mathbf{x}_1, t_1, \ldots \mathbf{x}_n, t_n) \equiv \langle A_1(\mathbf{x}_1, t_1) \cdots A_n(\mathbf{x}_n, t_n)\rangle^T$$

using an inductive formula linking truncated correlation functions to regular correlation functions:

$$w_n(\mathbf{x}_1, \tau_1, k_1 \ldots, \mathbf{x}_n, \tau_n, k_n) = \sum_{s=1}^{n} \sum_{\rho \in R_s} w_{\alpha_1}^T(\pi_1) \ldots w_{\alpha_s}^T(\pi_s).$$

Here R_s denotes the set of all partitions of the set $\{1, ..., n\}$ into subsets s denoted by $\pi_1, ..., \pi_s$, the number of elements in the subset π_i is denoted by α_i, and $w_{\alpha_i}^T(\pi_i)$ denotes the truncated correlation function with arguments $\mathbf{x}_a, \tau_a, k_a$, where $a \in \pi_i$. This formula expresses correlation functions in terms of truncated functions for all possible partitions of the set of indices; it can be used also to express truncated correlation functions in terms of usual ones.

When there are only two operators, the truncated correlation function has the form

$$w_2^T(\mathbf{x}_1, \tau_1, k_1, \mathbf{x}_2, \tau_2, k_2) = \omega(A_1(\mathbf{x}_1, t_1)A_2(\mathbf{x}_2, t_2)) - \omega(A_1(\mathbf{x}_1, t_1))\omega(A_2(\mathbf{x}_2, t_2)).$$

Since ω is translation-invariant and stationary, both usual and truncated correlation functions depend only on the differences $\mathbf{x}_i - \mathbf{x}_j, t_i - t_j$. We say that the cluster property is satisfied if the truncated correlation functions become small at $\mathbf{x}_i - \mathbf{x}_j \to \infty$. Smallness can be understood in different ways; we impose the strongest condition: at fixed t_i they tend to zero faster than any power of the difference $d = \min \|\mathbf{x}_i - \mathbf{x}_j\|$. More precisely, we assume that

$$|w_n^T(\mathbf{x}_1, t_1, \ldots \mathbf{x}_n, t_n)| \leq \frac{C_s(t)}{d^s},$$

where s is any natural number, and $C_s(t)$ is a polynomial function of times t_i.

In the simplest form, the cluster property means that

$$\omega(A(\mathbf{x}, t)B) = \omega(A)\omega(B) + \rho(\mathbf{x}, t),$$

where $\rho(\mathbf{x}, t)$ is small for large $\mathbf{x}$ (we used this property in Sect. 2.2).

We can go to momentum representation by applying the Fourier transform with respect to spatial variables. The invariance with respect to spatial translations leads to the appearance of δ-function of the sum of momenta $\mathbf{p}_i$. It follows from the cluster property that the truncated correlation function in momentum representation is a smooth function of momenta multiplied by the δ-function:

$$\nu_n(\mathbf{p}_2, \dots, \mathbf{p}_n, t_1, \dots, t_n)\delta(\mathbf{p}_1 + \dots + \mathbf{p}_n).$$

(The Fourier transform of a fast-decreasing function is smooth.)

In relativistic local quantum theory, the cluster property is satisfied if the particle masses are bounded from below by a positive number (mass gap).

2.5 Green's Functions. Generalized Green's Functions

2.5.1 Green's Functions

In Sect. 2.4 we defined the correlation function of translation-invariant stationary state ω as $\omega(M)$ where $M = A_1(\mathbf{x}_1, t_1) \dots A_r(\mathbf{x}_r, t_r)$. In the definition of Green's function, we replace M with a product where the same factors are ordered by time in descending order. This is what is called chronological product. (It is not defined when some times coincide, but this will be irrelevant in our considerations.) We can say that Green's function is the average (=expectation value) of chronological product in the state ω. Equivalently, we can say that we are taking the expectation value of this product with respect to the vector θ corresponding to ω in the GNS construction.

We obtain the function

$$G_n = \omega(T(A_1(\mathbf{x}_1, t_1) \dots A_r(\mathbf{x}_r, t_r))) = \langle\theta|T(\hat{A}_1(\mathbf{x}_1, t_1) \dots \hat{A}_r(\mathbf{x}_r, t_r))|\theta\rangle,$$

which is called the Green's function in $(\mathbf{x}, t)$-representation (in coordinate representation).

As always, we can go to the momentum representation by taking the Fourier transform over $\mathbf{x}$. This will be what is called the $(\mathbf{p}, t)$-representation (momentum and time). One can also take the (inverse) Fourier transform with respect to the time variable; then we obtain Green's function in the $(\mathbf{p}, \epsilon)$-representation, where the arguments are momenta and energies. We will need all these representations.

Due to translational invariance, Green's function in the $(\mathbf{x}, t)$-representation depends on the differences $\mathbf{x}_i - \mathbf{x}_j, t_i - t_j$. Therefore, we have a factor $\delta(\mathbf{p}_1 + \cdots + \mathbf{p}_r)$ in the $(\mathbf{p}, t)$−representation; it corresponds to the momentum conservation law. In the $(\mathbf{p}, \epsilon)$-representation we have also the factor $\delta(\epsilon_1 + \cdots + \epsilon_r)$, corresponding to the energy conservation law.

Let us consider poles of Green's function in $(\mathbf{p}, \epsilon)$-representation. It should be noted that we always ignore δ-functions when talking about the poles. In particular, when Green's function includes only two operators (two-point Green's function), in the $(\mathbf{p}, \epsilon)$-representation we have two momenta and two energies as arguments, but δ-functions allow us to get the following expression for this function:

$$G(\mathbf{p}_1, \epsilon_1 | A, A')\delta(\mathbf{p}_1 + \mathbf{p}_2)\delta(\epsilon_1 + \epsilon_2).$$

The function $G(\mathbf{p}_1, \epsilon_1 | A, A')$ depends on the variable $\mathbf{p}_1$ and the variable ϵ_1; the poles of it with respect to energy at fixed momentum correspond to particles. These poles depend on momentum, and the corresponding function $\varepsilon(\mathbf{p})$ gives the dispersion law for particles (dependence of energy on momentum). These well-known facts can be easily deduced from the reasoning that will be used below.

2.5.2 Generalized Green's Functions

To define generalized Green's functions we take the chronological product of some operators, multiply it by the anti-chronological product of other operators, and take the average (expectation value) in some state ω:

$$G_{mn} = \omega(MN).$$

Here M is defined as a chronological product:

$$M = T(\hat{A}_1^*(\mathbf{x}_1, t_1) \ldots \hat{A}_n^*(\mathbf{x}_m, t_m)) = T(\hat{A}^*).$$

(the times are decreasing). N is defined as the anti-chronological product (times increase):

$$N = T^{opp}(\hat{B}_1(\mathbf{x}_1', t_1') \ldots \hat{B}_n(\mathbf{x}_n', t_n')) = T^{opp}(\hat{B}).$$

G_{mn} is the generalized Green's function in the state ω.

In the algebraic approach, we consider operators $\hat{A}_i, \hat{B}_j$ as elements of the algebra $\mathcal{A}$.

Recall that for any $*$-algebra $\mathcal{A}$ any element $\hat{B} \in \mathcal{A}$ specifies two operators in the space $\mathcal{L}$ of continuous linear functionals on $\mathcal{A}$. One of them transforms a linear functional $\omega(A)$ into the functional $\omega(A\hat{B})$, the second one transforms $\omega(A)$ into the functional $\omega(\hat{B}^* A)$. The first of these operators is denoted by the symbol B, the second operator is denoted $\tilde{B}$.

If $\hat{B} = \hat{B}_1 \hat{B}_2$ then $(B\omega)(C) = \omega(C\hat{B}_1\hat{B}_2)$, hence

$$(B_1 B_2)\omega = B_2(B_1\omega)$$

(the order is changed because $\hat{B}$ acts from the right.)

It follows that

$$(T(\tilde{A}B)\omega)(\alpha) = \omega\Big(T(\hat{A}^*)\alpha T^{opp}(\hat{B})\Big) = \omega(M\alpha N). \tag{2.40}$$

Taking $\alpha = 1$ we obtain that generalized Green's functions can be defined in terms of chronological product: if $\hat{A}_1, ..., \hat{A}_m, \hat{B}_1, ..., \hat{B}_n \in \mathcal{A}$ the generalized Green's function G_{mn} can be represented by the formula

$$\langle 1|T\big(\tilde{A}_1(\mathbf{x}_1, t_1)...\tilde{A}_m(\mathbf{x}_m, t_m)B_1(\mathbf{x}'_1, t'_1)...B(\mathbf{x}'_n, t'_n)\big)|\omega\rangle. \tag{2.41}$$

(We are using bra-ket notations where bra-vectors are elements of $\mathcal{A}$ and ket vectors are elements of the dual space $\mathcal{L}$. In particular, for a linear functional σ on $\mathcal{A}$ and $A \in \mathcal{A}$ we have $\langle 1|A|\sigma\rangle = \sigma(A)$.)

The representation (2.41) will be used in Sect. 2.8.4 to construct perturbation theory for generalized Green's functions and in Sect. 4.8 to obtain diagram techniques for the calculation of them in cases when $\mathcal{A}$ is Weyl or Clifford algebra.

Generalized Green's functions play an important role in statistical physics (they appeared for the first time in Keldysh formalism of non-equilibrium statistical physics, see, for example, [6]). One can express various physical quantities (in particular, response functions) in terms of these functions.

Notice that the reasoning above prompts further generalization of Green's functions: for every element $\alpha \in \mathcal{A}$ and linear functional $\omega \in \mathcal{L}$ we define GGreen function G_{mn} by the formula

$$\langle\alpha|T\big(\tilde{A}_1(\mathbf{x}_1, t_1)...\tilde{A}_m(\mathbf{x}_m, t_m)B_1(\mathbf{x}'_1, t'_1)...B(\mathbf{x}'_n, t'_n)\big)|\omega\rangle. \tag{2.42}$$

It follows from (2.40) that GGreen function can be represented by the formula $\omega(M\alpha N)$.

More generally, in (2.42) α can belong to the space $\mathcal{L}^\vee$ dual to $\mathcal{L}$.

2.6 Scattering Amplitudes in Terms of Green's Functions. LSZ Formula

2.6.1 Introduction

Let us consider the definition of scattering in the algebraic approach. The consideration in Sect. 2.3 is based on axioms, which are not easy to check. Now we will impose requirements, that are much easier to check. In particular, they are satisfied in relativistic local theory. As before, our starting point is an associative algebra $\mathcal{A}$ with involution ($*$-algebra). Space-time translations are automorphisms of this algebra.

We will work with non-normalized states represented by positive linear functionals on $\mathcal{A}$; the cone of non-normalized states is denoted by C; a translation-invariant stationary state is denoted by $\omega \in C$. Excitations of ω states are elements of the pre-Hilbert space $\mathcal{H}$, which is constructed from ω using the GNS construction.

The algebra $\mathcal{A}$ is represented in the pre-Hilbert space $\mathcal{H}$ and its completion, the Hilbert space $\bar{\mathcal{H}}$, an operator corresponding to an element $A \in \mathcal{A}$ is denoted by $\hat{A}$. One can consider a normed algebra $\mathcal{A}(\omega)$ consisting of bounded operators $\hat{A}$.

Let us take an element A of the algebra $\mathcal{A}$ which is represented by a bounded operator $\hat{A}$ in Hilbert space $\bar{\mathcal{H}}$. We can consider temporal and spatial translations of this operator. The result will be denoted by $\hat{A}(\mathbf{x}, \tau)$.

We will consider asymptotically commutative algebras. In other words, we will require that the commutator of a shifted operator with another operator becomes small at large spatial shifts. This can be formalized in different ways. We will impose the condition

$$||[\hat{A}_1(\mathbf{x}, \tau), \hat{A}_2]|| \leq \frac{C_n(\tau)}{1 + ||\mathbf{x}||^n},$$

where $C_n(\tau)$ is a polynomial and n is arbitrary (strong asymptotic commutativity).

Replacing in this formula commutators with anticommutators we obtain the definition of asymptotic anticommutativity of two operators. We say that an algebra is asymptotically anicommutative if its generators asymptotically anticommute.

One can apply the arguments used in Sect. 2.3.2 to derive the existence of Møller matrices from strong asymptotic (anti)commutativity. On the other side, the existence of Møllller matrices can be derived from strong cluster property.

In what follows we talk about asymptotically commutative algebras, the case asymptotically anticommutative algebras is very similar.

Asymptotic commutativity and cluster property are related; under certain conditions, one can derive cluster property from asymptotic commutativity (see [25]). The proofs are more transparent in the framework of asymptotic commutativity therefore we will use cluster property only to prove that Møller matrices are isometric.

We can average an operator $\hat{A} \in \mathcal{A}(\omega)$ with a smooth and fast decreasing function $\alpha(\mathbf{x}, \tau)$:

$$B = \int d\tau d\mathbf{x} \alpha(\mathbf{x}, \tau) \hat{A}(\mathbf{x}, \tau). \tag{2.43}$$

It is possible to shift the operator B in time and space:

$$B(\mathbf{x}, \tau) = \int d\tau', d\mathbf{x}' \alpha(\mathbf{x} - \mathbf{x}', \tau - \tau') \hat{A}(\mathbf{x}', \tau').$$

One can differentiate under the sign of the integral. Since the function $\alpha(\mathbf{x}, \tau)$ is assumed to be smooth, $B(\mathbf{x}, \tau)$ is infinitely differentiable. We say that an operator of the form (2.43) is smooth; we always work with smooth operators.

It is easy to check that smooth operators and their derivatives asymptotically commute if they were constructed from elements of asymptotically commutative algebra. It follows that (2.28) is correct for smooth operator B and non-overlapping functions f_i, f_j. This allows us to use the considerations of the preceding section to prove the existence of Møller matrices and to define the scattering matrix.

2.6.2 LSZ Formula

We will prove that to find the scattering amplitudes one should consider the asymptotic behavior of Green's function in $(\mathbf{p}, t)$-representation when $t \to \pm\infty$. This observation and the remark that asymptotic behavior in the $(\mathbf{p}, t)$-representation is governed by the poles in the $(\mathbf{p}, \epsilon)$-representation allow us to say that scattering amplitudes can be expressed in terms of residues in these poles (in terms of on-shell values of Green's functions). This is the Lehmann, Symanzik, and Zimmermann (LSZ) formula.

We use a well-known mathematical fact: If the asymptotic behavior of a function $\rho(t)$ at $t \to \pm\infty$ has the form $e^{-itE_\pm}A_\pm$ or, put another way, there is a limit $\lim_{t\to\pm\infty} e^{itE_\pm}\rho(t) = A_\pm$, then the (inverse) Fourier transform $\rho(\epsilon)$ has poles at the points $E_\pm \pm i0$ with residues $\mp 2\pi i A_\pm$. In other words, the limit corresponds to the residues in the poles and the exponents correspond to poles; the poles are slightly shifted in the complex plane either up or down from the real axis. This fact is very important for us.

Below we will prove the LSZ formula under certain conditions. First of all, we assume that the theory has an interpretation in terms of particles. This means that the Møller matrices $S_\pm$ are unitary. Both S_- and S_+ give unitary equivalence between the free Hamiltonian in the asymptotic space $\mathcal{H}_{as}$ and the Hamiltonian in the space $\mathcal{H}$ obtained with the GNS procedure. Second, we assume that the conservation laws for energy and momentum guarantee the stability of particles.

To simplify the notation, we will discuss the case when there is only one type of particles. (The same considerations are valid if we have several particles with the same dispersion law as it happens for particles with spin.) Recall that we considered a generalized function $\Phi(\mathbf{p})$ corresponding to the state of a particle with a given momentum $\mathbf{p}$, and this state is an eigenvector for both momentum and energy operators. The Hamiltonian acts on $\Phi(\mathbf{p})$ as multiplication by the function $\varepsilon(\mathbf{p})$ (dispersion law):

$$\hat{H}\Phi(\mathbf{p}) = \varepsilon(\mathbf{p})\Phi(\mathbf{p}), \quad \hat{\mathbf{P}}\Phi(\mathbf{p}) = \mathbf{p}\Phi(\mathbf{p}).$$

We must remember that $\Phi(\mathbf{p})$ is a generalized function. We should integrate it with some test function $\phi(\mathbf{p})$ to get a vector $\Phi(\phi) = \int d\mathbf{p}\phi(\mathbf{p})\Phi(\mathbf{p})$.

Now we want to assume that the one-particle spectrum does not overlap with the multi-particle spectrum.

Let us formulate this assumption more precisely. Let us denote by $\mathcal{H}_0$ the one dimensional subspace containing vector θ, by $\mathcal{H}_1$ the smallest closed subspace of $\mathcal{H}$

containing all vectors $\Phi(\phi)$ (one-particle space) and by $\mathcal{H}_{\mathcal{M}}$ the orthogonal complement of the direct sum $\mathcal{H}_0 + \mathcal{H}_1$ (multiparticle space).

We assume that the joint spectra of the Hamiltonian and momentum operator in these three spaces do not overlap.

The asymptotic Hamiltonian is free. It (and hence $\hat{H}$, that is unitary equivalent to the asymptotic Hamiltonian) has a spectrum completely determined by the function $\varepsilon(\mathbf{p})$. The energies of multiparticle excitations are simply the sums $\varepsilon(\mathbf{p}_1) + \cdots + \varepsilon(\mathbf{p}_n)$, corresponding momenta are $\mathbf{p}_1 + ... + \mathbf{p}_n$ (here $n > 1$). To say that the one-particle spectrum does not overlap with the multi-particle spectrum, we must require the inequality

$$\varepsilon(\mathbf{p}_1 + ... + \mathbf{p}_n) \neq \varepsilon(\mathbf{p}_1) + ... + \varepsilon(\mathbf{p}_n).$$

This means that a particle with momentum $\mathbf{p}_1 + ... + \mathbf{p}_n$ cannot decay into particles with momenta $\mathbf{p}_1, ..., \mathbf{p}_n$. The conservation laws for momentum and energy forbid decay.

Notice that the statements below can be proved also in more general situations when the decay is forbidden by some conservation laws (one should assume that in addition to momentum and energy, we have some other integrals of motion).

Let us formulate the LSZ formula. To do this we will fix some elements $A_i \in \mathcal{A}$ of the algebra $\mathcal{A}$. We require that by applying the operator $\hat{A}_i$ to the vector θ (which in relativistic quantum theory is interpreted as the physical vacuum) and projecting on one-particle space we get a non-zero vector. More precisely, the projection of the vector $\hat{A}_i\theta$ should be a one-particle state of the form:

$$\Phi(\phi_i) = \int \phi_i(\mathbf{p})\Phi(\mathbf{p})d\mathbf{p},$$

where $\phi_i(\mathbf{p})$ is a function which does vanish anywhere. The projection of this vector onto the vector θ must vanish.

Let us consider Green's functions defined as expectation values of chronological products of the elements A_i and their adjoint elements A_i^*. We take Green's function in $(\mathbf{x}, t)$-representation:

$$G_{mn} = \omega(T(A_1^*(\mathbf{x}_1, t_1) \dots A_m^*(\mathbf{x}_m, t_m)A_{m+1}(\mathbf{x}_{m+1}, t_{m+1}) \dots A_{m+n}(\mathbf{x}_{m+n}, t_{m+n}))).$$

Then we go to $(\mathbf{p}, t)$- and $(\mathbf{p}, \epsilon)$-representation. It is convenient to change the sign of the variables $\mathbf{p}_i$ and ϵ_i for $1 \leq i \leq m$. We multiply the Green's function in the $(\mathbf{p}, \epsilon)$- representation by the expression:

$$\prod_{1\leq i\leq m} \overline{\Lambda_i(\mathbf{p}_i)}(\epsilon_i + \varepsilon(\mathbf{p}_i)) \prod_{m<j\leq m+n} \Lambda_j(\mathbf{p}_j)(\epsilon_j - \varepsilon(\mathbf{p}_j)).$$

where we introduced the notation $\Lambda_i(\mathbf{p}) = \phi_i(\mathbf{p})^{-1}$.

Then we take the limit $\epsilon_i \to -\varepsilon(\mathbf{p}_i)$ for $1 \leq i \leq m$ and $\epsilon_j \to \varepsilon(\mathbf{p}_j)$ for $m < j \leq m + n$.

Only poles will contribute to the limit. In other words, the calculation boils down to taking residues of the poles.

We can do this procedure in two steps. First, we multiply the Green's function by

$$\prod_{1\leq i\leq m} (\epsilon_i + \varepsilon(\mathbf{p}_i)) \prod_{m<j\leq m+n} \epsilon_j - \varepsilon(\mathbf{p}_j)).$$

and take the limit $\epsilon_i \to -\varepsilon(\mathbf{p}_i)$ for $1 \leq i \leq m$ and $\epsilon_j \to \varepsilon(\mathbf{p}_j)$ for $m < j \leq m+n$. At the end we multiply by $\Lambda_i(\mathbf{p})$ for $i > m$ and by $\overline{\Lambda_i(\mathbf{p})}$ if $i \leq m$. In quantum field theory, this is called the renormalization of the wave function. In the case when these factors are not included, we talk about on-shell Green's functions, and if included, we say that we consider normalized on-shell Green's functions.

The basic statement in the approach of Lehmann, Symanzik and Zimmermann is that the normalized on-shell Green's function gives the scattering amplitude (LSZ formula).

2.6.3 *Proof of LSZ Formula*

To prove LSZ formula, we will first consider the case where the operators $\hat{A}_i$ give one-particle states $\hat{A}_i\theta = \Phi(\phi_i)$ (no need to project). We will call them good operators. At the end of the section, we will explain that the general case can be reduced to this particular case.

So far we considered the case when there is only one type of particles. Let us consider the case when there are many types of particles, in other words, there are many functions $\Phi_k(\mathbf{p})$ which are eigenvectors for both momentum and energy:

$$\mathbf{P}\Phi_k = \mathbf{p}\Phi_k(\mathbf{p}), \quad H\Phi_k(\mathbf{p}) = \varepsilon_k(\mathbf{p})\Phi_k(\mathbf{p}),$$

(The dispersion laws $\varepsilon_k(\mathbf{p})$ are given by smooth functions. They can coincide, for example, when we are dealing with particles with spin, but in general, they are different.) As always, $\Phi_k(\mathbf{p})$ are generalized functions, i.e., we should integrate them with test functions to get vectors from $\mathcal{H}$. We consider test functions from the space $\mathcal{S}$ of smooth fast-decreasing functions. To guarantee that time shifts are well-defined in the space $\mathcal{S}$, we should assume that the functions $\varepsilon_k(\mathbf{p})$ grow at most polynomially.

As already mentioned, we will work with good operators $B_k \in \mathcal{A}$ (operators which are smooth and transform vector θ into one-particle states $\hat{B}_k\theta = \Phi_k(\phi_k)$). Now we define the operator $\hat{B}_k(f,t)$ depending on the function $f = f(\mathbf{p})$ as follows:

$$\hat{B}_k(f,t) = \int \tilde{f}(\mathbf{x},t)\hat{B}_k(\mathbf{x},t)d\mathbf{x},$$

where the function $\tilde{f}(\mathbf{x}, t)$ is obtained as the Fourier transform of the function $f(\mathbf{p})e^{-i\varepsilon_k(\mathbf{p})t}$ with respect to the momentum variable.

Similar operators were considered in Sect. 2.3.1. They have the property that applying them to θ we obtain a t-independent one-particle state

$$\hat{B}_k(f, t)\theta = \Phi_k(f\phi_k)$$

hence

$$\dot{\hat{B}}_k(f, t)\theta = 0, \tag{2.44}$$

where the dot stands for the time derivative. (In Sect. 2.3.1, the function ϕ_k was equal to 1.) The fact that the resulting state is independent of time is the result of a formal calculation. The calculations become quite simple if we introduce operators

$$\hat{B}_k(\mathbf{p}, t) = \int d\mathbf{x}e^{-i\mathbf{p}\mathbf{x}}\hat{B}_k(\mathbf{x}, t).$$

Now we repeat the considerations of Sect. 2.3.2, but the notations have changed because now it is more convenient to write the indices explicitly.

We introduce a vector

$$\Psi(k_1, f_1, \ldots, k_n, f_n|t_1, \ldots, t_n) = \hat{B}_{k_1}(f_1, t_1)\cdots\hat{B}_{k_n}(f_n, t_n)\theta, \tag{2.45}$$

where it is assumed that the functions $f_1, \ldots, f_n$ have compact supports.

Now, as in Appendix A.3, we consider vectors $\mathbf{v}_i(\mathbf{p}) = \nabla\varepsilon_{k_i}(\mathbf{p})$, which can be interpreted as velocities. We denote by U_i an open set containing all possible velocities $\mathbf{v}_i(\mathbf{p})$ where $\mathbf{p}$ belongs to the support of the function f_i. We require that all these sets do not overlap, then we will call the functions f_1, ..., f_n non-overlapping. This means that all classical velocities are different and therefore the wave packets are moving in different directions. Then in the coordinate representation, the corresponding wave functions almost do not overlap (essential supports do not overlap).

Now we will take the limit $t_i \to \infty$. One can prove that the vector $\Psi(k_1, f_1, \ldots, k_n, f_n|t_1, \ldots, t_n)$ has a limit, denoted by

$$\Psi(k_1, f_1, \ldots, k_n, f_n| \pm \infty).$$

For simplicity, we give the proof in the case $t_i = t$ (all times coincide). It is sufficient to prove that the derivative with respect to t is a summable function. By definition, to get the vector Ψ we repeatedly apply operators $\hat{B}_{k_i}(f_i, t_i)$ to θ. When we differentiate the expression for Ψ with respect to t, we have a dot (denoting the time derivative) over one of the operators $\hat{B}$. We can move the operator with a dot to the right using the asymptotic commutativity and (A.6) (but only if we work with non-overlapping functions). We get additional summands that are summable functions of t. The operator with a dot applied to θ gives zero due to (2.44), hence the derivative with respect to t is summable. This means that there is a limit.

Since the limit exists, we can define Møller matrices. To do this, we introduce the asymptotic space $\mathcal{H}_{as}$ as a Fock representation of the operators $a_k^*(f), a_k(f)$ and define the Møller matrices S_- and S_+ as operators defined on a subset of $\mathcal{H}_{as}$ and taking values in $\bar{\mathcal{H}}$ by the formula:

$$\Psi(k_1, f_1, \dots, k_n, f_n| \pm \infty) = S_\pm(a_{k_1}^*(f_1\phi_{k_1}) \dots a_{k_n}^*(f_n\phi_{k_n})|0\rangle),$$

where $|0\rangle$ is the Fock vacuum. This is the same formula as in Sect. 2.3.2 with the difference that now we have factors ϕ_{k_i}. (Recall that a good operator $\hat{B}_{k_i}$ acting on θ gives $\Phi_{k_i}(\phi_{k_i})$.)The Møller matrices are defined on a dense subspace of the asymptotic Hilbert space $\mathcal{H}_{as}$.

The physical meaning of Møller matrices can be understood from the following formula (which in other notations was written in Sect. 2.3.2):

$$e^{-iHt}\Psi(k_1, f_1, \dots, k_n, f_n| \pm \infty) = S_\pm(a_{k_1}^*(f_1\phi_{k_1}e^{-i\varepsilon_{k_1}t}) \dots a_{k_n}^*(f_n\phi_{k_n}e^{-i\varepsilon_{k_n}t})|0\rangle).$$

This formula means that when we consider evolution in the space $\bar{\mathcal{H}}$, in the limit $t \to \pm\infty$ the action of the evolution operator on the vector Ψ corresponds in the asymptotic space to the evolution governed by a free Hamiltonian. We can say that the evolution of the vector Ψ for large t corresponds to the evolution of a system of n distant particles with non-overlapping wave functions $f_1\phi_{k_1}e^{-i\varepsilon_{k_1}t}, \dots, f_n\phi_{k_n}e^{-i\varepsilon_{k_n}t}$.

Our definition of $S_\pm$ can be ambiguous. For example, we can use different good operators and it is unclear whether we get the same answer. However, we can prove that the answer does not depend on our choices. We will derive from cluster property that $S_\pm$ are isometric operators. They preserve the norm and preserve the scalar product. Such operators cannot be multivalued. (If two vectors coincide, the distance between them is 0, hence two coinciding vectors must go to coinciding vectors.) At the same time we can see that the vector $\Psi(k_1, f_1, ..., k_n, f_n| \pm \infty)$ does not change when the arguments (k_i, f_i) and (k_j, f_j). are permuted.

Let us sketch the proof.

To define Møller matrices we used vectors $\Psi(k_1, f_1, \dots, k_n, f_n|t_1, \dots, t_n)$ specified by the formula (2.45). Note that according to (2.45) such a vector is obtained by repeatedly applying the operators B to θ. It is easy to see that, the scalar product of two such vectors can be expressed in terms of correlation functions defined as the average values (expectation values) of products of the operators B and B^*. The correlation functions are expressed in terms of truncated correlation functions; in truncated correlation functions only two-point correlation functions survive in the limit, $t \to \pm\infty$ if we require cluster property. This remark relates the scalar product of two vectors of the form $\Psi(k_1, f_1, \dots, k_n, f_n| \pm \infty)$ to the scalar product in the asymptotic space. This allows us to say that a Møller matrix is an isometric mapping.

Once we have introduced the notion of Møller matrix, we can introduce the notions of *out*-operators and *in*-operators:

$$a_{in}(f)S_- = S_- a(f), \quad a^*_{in}(f)S_- = S_- a^*(f),$$

$$a_{out}(f)S_+ = S_+ a(f), \quad a^*_{out}(f)S_+ = S_+ a^*(f).$$

(Again, we do not write an index describing the type of particles.)

It is easy to check that

$$a^*_{in}(f\phi) = \lim_{t\to-\infty} \hat{B}(f,t), \; a^*_{out}(f\phi) = \lim_{t\to\infty} \hat{B}(f,t).$$

The limit in (2.6.3) exists on the set of all vectors of the form $\Psi(k_1, f_1, \ldots, k_n, f_n|\pm)$ provided that f, f_1, ..., f_n-is a non-overlapping family of functions. Interestingly, when the dimension of the space $d \geq 3$, under some conditions this limit exists without the non-overlapping condition. (This is insignificant for us, because the non-overlapping condition gives a limit on a dense subset, and this is sufficient.)

Now, based on what we know, we can write out explicitly how the operators we have defined act on the in-states:

$$a^*_{in}(f\phi)\Psi(f_1, \ldots, f_n| -\infty) = \Psi(f, f_1, \ldots, f_n| -\infty),$$

$$a^*_{out}(f\phi)\Psi(f_1, \ldots, f_n|\infty) = \Psi(f, f_1, \ldots, f_n|\infty).$$

$$a_{in}(f)\Psi(\phi^{-1}f, f_1, \ldots, f_n| -\infty) = \Psi(f_1, \ldots, f_n| -\infty),$$

$$a_{out}(f)\Psi(\phi^{-1}f, f_1, \ldots, f_n|\infty) = \Psi(f_1, \ldots, f_n|\infty).$$

These formulas can be seen as definitions of operators a_{in} and a_{out}. Roughly speaking, the operators $a^*_{in/out}$ add one function to $\Psi(f_1, \ldots, f_n| \mp\infty)$, and the corresponding annihilation operators destroy one of these functions.

Recall, that we say that the theory has an interpretation in terms of particles if the operators S_+ and S_- are unitary. In this case (and in the more general case when the image of S_- coincides with the image of S_+), we can define the scattering matrix (S-matrix):

$$S = S^*_+ S_-.$$

as a unitary operator in the asymptotic space $\mathcal{H}_{as}$. The asymptotic space is a Fock space. It has a generalized basis $|\mathbf{p}_1, \ldots, \mathbf{p}_n\rangle = \frac{1}{n!}a^*(\mathbf{p}_1)\ldots a^*(\mathbf{p}_n)|0\rangle$.

In this basis the matrix elements of the unitary operator S (scattering amplitudes) can be expressed in in terms of in- and out-operators. We get the following formula:

$$S_{mn}(\mathbf{p}_1, \ldots, \mathbf{p}_m|\mathbf{q}_1, \ldots, \mathbf{q}_n) = \langle a^*_{in}(\mathbf{q}_1)\ldots a^*_{in}(\mathbf{q}_n)\theta, a^*_{out}(\mathbf{p}_1)\ldots a^*_{out}(\mathbf{p}_m)\theta\rangle \tag{2.46}$$

This follows directly from the definition of in- and out-operators. In the formula (2.46) and in what follows, we omit some numerical coefficients.

The above formula is proved only for the case when all the momentum values $\mathbf{p}_i, \mathbf{p}_j$ are different. (More precisely, we must assume that all vectors $\mathbf{v}(\mathbf{p}_i) = \nabla\varepsilon(\mathbf{p}_i), \mathbf{v}(\mathbf{q}_j) = \nabla\varepsilon(\mathbf{q}_j)$ are different. When the function $\varepsilon(\mathbf{p})$ is strictly convex, it is sufficient to assume that $\mathbf{p}_i, \mathbf{q}_j$ are different.) The formula (2.46) should be understood in the sense of generalized functions. This means that the set of functions $f_i(\mathbf{p}_i), g_j(\mathbf{q}_j)$ with non-overlapping subsets $\overline{U(f_i)}, \overline{U(g_j)}$ should be taken as test functions. In other words, a rigorous formulation of (2.46) is the formula:

$$S_{mn}(f_1, \ldots, f_m | g_1, \ldots, g_n) =$$

$$= \int d^m\mathbf{p} d^n\mathbf{q} \prod f_i(\mathbf{p}_i) \prod g_j(\mathbf{q}_j) S_{mn}(\mathbf{p}_1, \ldots, \mathbf{p}_m | \mathbf{q}_1, \ldots, \mathbf{q}_n) =$$

$$= \langle a^*_{in}(g_1) \ldots a^*_{in}(g_n)\theta, a^*_{out}(\bar{f}_1) \ldots a^*_{out}(\bar{f}_m)\theta \rangle$$

where functions f_1, ..., f_m, g_1, ..., g_n do not overlap.

Recalling that we defined the S-matrix (scattering matrix) taking limits $t \to \pm\infty$ and using the formulas (2.6.3), we arrive at the following representation:

$$S_{mn}(f_1, \ldots, f_m | g_1, \ldots, g_n) =$$

$$\lim_{t\to\infty, \tau\to-\infty} \langle \theta | \hat{B}(\bar{f}_m\phi^{-1}, t)^* \ldots \hat{B}(\bar{f}_1\phi^{-1}, t)^* \hat{B}(g_1\phi^{-1}, \tau) \ldots \hat{B}(g_n\phi^{-1}, \tau)) | \theta \rangle =$$

$$\lim_{t\to\infty, \tau\to-\infty} \omega(B(\bar{f}_m\bar{\phi}^{-1}, t)^* \ldots B(\bar{f}_1\bar{\phi}^{-1}, t)^* B(g_1\phi^{-1}, \tau) \ldots B(g_n\phi^{-1}, \tau)),$$

where $B(f, t)^* = \int d\mathbf{x} B^*(\mathbf{x}, t)\overline{\tilde{f}(\mathbf{x}, t)}$.

We can also write a more general formula

$$S_{mn}(f_1, \ldots, f_m | g_1, \ldots, g_n) =$$

$$\lim_{\substack{t_i\to\infty,\\ \tau_j\to-\infty}} \omega(B_m(\bar{f}_m\phi_m^{-1}, t_m)^* \ldots B_1(\bar{f}_1\phi_1^{-1}, t_1)^* B_{m+1}(g_1\phi_{m+1}^{-1}, \tau_1) \ldots B_{m+n}(g_n\phi_{m+n}^{-1}, \tau_n)),$$

where all B_i are different good operators and $B_i\theta = \Phi(\phi_i)$.

A very important observation: it follows from the non-overlapping condition that in the limit $t_i \to \infty$, $\tau_j \to -\infty$ the order of factors is irrelevant both in the group with times tending to +infinity and in the group with times tending to—infinity. This means that for large times we can rearrange these operators, in particular, we can them by time. This means that we can regard the expression under the sign of limit as Green's function. This is the end of the proof of the statement that the matrix element of the scattering matrix can be expressed in terms of the asymptotic behavior of Green's function.

We can express the operators $\hat{B}_k(f,t)$ in terms of $\hat{B}_k(\mathbf{p},t)$ and get the following result:

$$S_{mn}(f_1,\ldots,f_m|g_1,\ldots,g_n)=$$

$$\int d^{m+n}\mathbf{p}\lim_{\substack{t_i\to\infty,\\ \tau_j\to-\infty}}\langle\theta|f_m\bar{\phi}_m^{-1}e^{i\varepsilon_m(\mathbf{p}_m)t_m}\hat{B}_m(\mathbf{p}_m,t_m)^*\ldots f_1\bar{\phi}_1^{-1}e^{i\varepsilon_1(\mathbf{p}_1)t_1}\hat{B}_1(\mathbf{p}_1,t_1)^*\times$$

$$g_1\phi_{m+1}^{-1}e^{-i\varepsilon_{m+1}(\mathbf{p}_{m+1})\tau_1}\hat{B}_{m+1}(\mathbf{p}_{m+1},\tau_n)\ldots g_n\phi_{m+n}^{-1}e^{-i\varepsilon_{m+n}(\mathbf{p}_{m+n})\tau_n}\hat{B}_{m+n}(\mathbf{p}_{m+n},\tau_n)|\theta\rangle.$$

We obtain the following formula for the matrix elements of the scattering matrix:

$$S_{mn}(\mathbf{p}_1,\ldots,\mathbf{p}_m|\mathbf{p}_{m+1}\ldots,\mathbf{p}_{m+n})=$$

$$\lim_{\substack{t_1,\ldots,t_m\to\infty,\\ t_{m+1},\ldots,t_{m+n}\to-\infty}}\langle\theta|\bar{\phi}_m^{-1}e^{i\varepsilon_m(\mathbf{p}_m)t_m}\hat{B}_m(\mathbf{p}_m,t_m)^*\ldots\bar{\phi}_1^{-1}e^{i\varepsilon_1(\mathbf{p}_1)t_1}\hat{B}_1(\mathbf{p}_1,t_1)^*\times$$

$$\phi_{m+1}^{-1}e^{-i\varepsilon_{m+1}(\mathbf{p}_{m+1})t_{m+1}}\hat{B}_{m+1}(\mathbf{p}_{m+1},t_{m+1})\ldots\times$$

$$\phi_{m+n}^{-1}e^{-i\varepsilon_{m+n}(\mathbf{p}_{m+n}),t_{m+n}}\hat{B}_{m+n}(\mathbf{p}_{m+n},t_{m+n})|\theta\rangle,$$

This formula tells us that starting with good operators, we can express the scattering matrix in terms of Green's functions in $(\mathbf{p},t)$-representation, or, more precisely, in terms of their asymptotics at $t_i\to\infty$ for $i\le m$ and $t_j\to-\infty$ for $i>m$. Factors ϕ_i^{-1} coincide with factors Λ_i that were introduced to get normalized Green's functions. The fact that we are considering asymptotic behavior in $(\mathbf{p},t)$ representation means that we are considering on-shell Green's functions in the energy representation. The fact that we get factors ϕ_i^{-1} means that we get normalized Green's functions.

We gave proof of LSZ formula for good operators. From it, one can conclude that it is true for a much broader class of operators. Let us show that in the approach of Lehmann, Symanzik and Zimmermann the operators A_i are almost arbitrary, assuming for simplicity that there exists only one type of particles. However, it is necessary that the projection of the vector $\hat{A}_i\theta$ on the one-particle space is nonzero, and the projection of this vector on the vector θ vanishes.

It is important that in the definition of the on-shell Green's function, the operators A_i can be replaced by smooth operators $A_i'=\int\alpha(\mathbf{x},t)A_i(\mathbf{x},t)d\mathbf{x}dt$. It is easy to check that this does not change normalized on-shell Green's functions. The proof is based on the remark that $A_(' \mathbf{x},t)$ can be obtained as a convolution of $\alpha(\mathbf{x},t)$ with $A_i(\mathbf{x},t)$. This is the first observation. And the second observation is that for an appropriate choice of α one can consider A_i' as good operators. Namely, we can take $\alpha(\mathbf{x},t)$ in such a way that the support of its Fourier transform $\hat{\alpha}(\mathbf{p},\omega)$ does not intersect with the multiparticle spectrum and does not contain zero. (We assumed that the one-particle spectrum does not intersect the multiparticle spectrum.) In this case, we automatically obtain a good operator.

Let us sketch the proof of this fact. We know that the operator $A'_i\mathbf{x}, t)$ is obtained from $A_i(\mathbf{x}, t)$ by convolution with $\alpha(\mathbf{x}, t)$. In the $(\mathbf{p}, \epsilon)$-representation the convolution turns into multiplication by the Fourier transform $\hat{\alpha}(\mathbf{p}, \omega)$ of $\alpha(\mathbf{x}, t)$. If we consider the spectrum of the energy and momentum operators, multiplication by the function $\hat{\alpha}(\mathbf{p}, \omega)$ in the $(\mathbf{p}, \epsilon)$-representation kills all points of the spectrum where this function is equal to zero. The function we consider kills the multi-particle spectrum, and we get a good operator.

Another possibility to prove the LSZ formula for a large class of operators is to consider admissible operators in the sense of Sect. 2.3.4 instead of good operators in the proof.

2.7 Inclusive Scattering Matrix

Now we want to apply these results to express the inclusive scattering matrix in terms of generalized Green's functions. For this purpose, we will consider in-states. We considered them as vectors in Hilbert space, but now we will consider them as positive functionals on algebra. Every vector $\Psi \in \bar{\mathcal{H}}$ specifies a positive functional on the algebra. If we apply some operator, for example $B(f, \tau)$ to a vector Ψ, then to get a positive functional corresponding to $B(f, \tau)\Psi$ we should apply an operator $L(f, \tau) = \tilde{B}(f, \tau)B(f, \tau)$ to the positive functional corresponding to Ψ. (Recall that the algebra element defines two operators on functionals: we can either multiply the argument from the right or multiply the argument by the adjoint element from the left.)

In terms of such operators, the in-state corresponding to the vector (2.21) and regarded as a positive functional can be written in the following form:

$$\nu = \lim_{\tau \to -\infty} L(f_1, \tau)...L(f_n, \tau)\omega.$$

Now consider the following expression:

$$\langle 1|L(f'_1, \tau')...L(f'_{n'}, \tau')L(f_1, \tau)...L(f_n, \tau)|\omega\rangle.$$

One can prove the existence of a limit of this expression as at $\tau' \to +\infty, \tau \to -\infty$ assuming that all functions f'_i and f_j do not overlap. (Recall that asymptotic commutativity is always assumed.) We denote this limit by Q; it can be represented in the form

$$Q = \lim_{\tau' \to +\infty} \langle 1|L(f'_1, \tau')...L(f'_{n'}, \tau')|\nu\rangle$$

Expressing Q in terms of functions g_k, g'_k related to functions $f_k, f'k$ by the formula (2.37) we obtain an expression $Q = Q(g'_1, ..., g'_{n'}, g_1, ..., g_n)$ that will be called the inclusive scattering matrix. Inclusive cross-sections can be calculated in terms of the inclusive scattering matrix. This is clear from the expression

$$Q = \nu(a_{out}(g'_{n'})...a_{out}(g'_1)a^*_{out}(g'_1)...a^*_{out}(g'_{n'}))).$$

that follows the relation $\langle 1|\sigma\rangle = \sigma(1)$ and the formula $\lim_{\tau'\to+\infty} B(f'_k, \tau') = a^*_{out}(g'_k)$ (see (2.38)). (Recall that the notions of inclusive cross-section and inclusive scattering matrix were discussed in Sect. 2.3.4.)

Q is a non-linear expression with respect to g and g'; it is a quadratic (or rather hermitian) expression. Any quadratic expression can be extended to a bilinear form, and a hermitian expression can be extended to a sesquilinear form—for one variable it will be linear, for another it will be antilinear.

To obtain an expression that will be linear (or somewhere antilinear), we introduce the notation $L(\tilde{g}, g, \tau) = \tilde{B}(\tilde{f}, \tau)B(f, \tau)$. The variables f and $\tilde{f}$ are separated here, and now what used to be treated as a hermitian expression will be treated as a sesquilinear expression depending on a doubled number of variables:

$$\rho(\tilde{g}'_1, g'_1, ..., \tilde{g}'_{n'}, g'_{n'}, \tilde{g}_1, g_1, ..., \tilde{g}_n, g_n) =$$

$$\lim_{\substack{\tau'\to+\infty,\\ \tau'\to-\infty}} \langle 1|L(\tilde{g}'_1, g'_1, \tau')...L(\tilde{g}'_{n'}, g'_{n'}, \tau')L(\tilde{g}_1, g_1, \tau)...L(\tilde{g}_n, g_n, \tau)|\omega\rangle.$$

This expression will also be called the inclusive scattering matrix and we will express it in terms of generalized Green's functions. First of all, we represent it in the form

$$\rho(\tilde{g}'_1, g'_1, ..., \tilde{g}'_{n'}, g'_{n'}, \tilde{g}_1, g_1, ..., \tilde{g}_n, g_n) =$$

$$\lim_{\substack{\tau'\to+\infty,\\ \tau\to-\infty}} \langle 1|B(f'_1, \tau')...B(f'_{n'}, \tau')\tilde{B}(\tilde{f}'_{n'}, \tau')\tilde{B}(\tilde{f}'_1, \tau') \times$$

$$B(f_1, \tau)...B(f_n, \tau)\tilde{B}(\tilde{f}_n, \tau)...\tilde{B}(\tilde{f}_1, \tau)|\omega\rangle.$$

It is convenient to use a more general formula

$$\rho(\tilde{g}'_1, g'_1, ..., \tilde{g}'_{n'}, g'_{n'}, \tilde{g}_1, g_1, ..., \tilde{g}_n, g_n) =$$

$$\lim_{\substack{\tau'_j, \tilde{\tau}'_j\to+\infty,\\ \tau_i, \tilde{\tau}_i\to-\infty}} \langle 1|B(f'_1, \tau'_1)...B(f'_{n'}, \tau'_{n'})\tilde{B}(\tilde{f}'_{n'}, \tilde{\tau}'_{n'})\tilde{B}(\tilde{f}'_1, \tilde{\tau}'_1) \times$$

$$B(f_1, \tau_1)...B(f_n, \tau_n)\tilde{B}(\tilde{f}_n, \tilde{\tau}_n)...\tilde{B}(\tilde{f}_1, \tilde{\tau}_1)|\omega\rangle.$$

We can assume that in this expression the times are ordered. (Part of times tends to $+\infty$, another part to $-\infty$. Within each group, due to asymptotic commutativity, we can rearrange the factors in any order, in particular, in order of decreasing times.) Applying the formula (2.40) we see that what stands under the limit sign can be expressed in terms of generalized Green's functions. The inclusive scattering matrix is expressed in terms of asymptotic behavior of these functions when times tend to $\pm\infty$. The same reasoning was used for the usual scattering matrix, only the number of arguments doubled.

2.8 Perturbation Theory. Feynman Diagrams

2.8.1 Perturbation Theory for Evolution Operator

Let us consider an equation $\frac{d\sigma}{dt} = H(t)\sigma$ in Banach (or topological linear) space. The evolution operator transforming $\sigma(t_0)$ into $\sigma(t)$ is denoted $U(t, t_0)$, it obeys

$$\frac{dU(t, t_0)}{dt} = H(t)U(t, t_0), U(t_0, t_0) = 1. \tag{2.47}$$

Here $H(t)$ can be interpreted either as a Hamiltonian of conventional quantum mechanics multiplied by $-i$ or as a "Hamiltonian" of geometric approach.

Let us assume that $H(t) = H(0) + gh(t)V$ (it is convenient to say that $H(0)$ is a free Hamiltonian and the rest is an interaction). Our goal is to represent $U(t, t_0|g$ as a formal power series with respect to g:

$U(t, t_0|g) = \sum g^n U_n(t, t_0)$.

Notice that it follows from (2.47) that the operator $S(t, t_0|g) = e^{-H(0)t}U(t, t_0|g)e^{H(0)t_0}$ (the evolution operator in the interaction picture) obeys

$$\frac{dS(t, t_0)|g}{dt} = gh(t)\mathrm{V}(t)S(t, t_0|g), S(t_0, t_0|g) = 1 \tag{2.48}$$

where $\mathrm{V}(t) = e^{-H(0)t}Ve^{H(0)t}$. Introducing the notation

$$S(t, t_0|g) = \sum g^n S_n(t, t_0)$$

we obtain

$\frac{dS_n(t,t_0)}{dt} = h(t)\mathrm{V}(t)S_{n-1}(t, t_0)$, $S_0(t_0, t_0) = 1$, $S_n(t_0, t_0) = 0$ when $n > 0$, hence

$$S_n(t, t_0) = \int_{t_0}^{t} d\tau h(\tau)\mathrm{V}(\tau)S_{n-1}(\tau, t_0).$$

It follows that

$$S_n(t, t_0) = \int_{t\geq\tau_n\geq\tau_{n-1}\geq\ldots\geq\tau_1\geq t_0} d\tau_n\ldots d\tau_1 h(\tau_n)\ldots h(\tau_1)\mathrm{V}(\tau_n)\ldots\mathrm{V}(\tau_1). \tag{2.49}$$

In terms of chronological product (times decreasing), we can present (2.49) as

$$S_n(t, t_0) = \frac{1}{n!} \int_{t \geq \tau_i \geq t_0} d\tau_n ... d\tau_1 h(\tau_n) ... h(\tau_1) T(\mathrm{V}(\tau_n) ... \mathrm{V}(\tau_1)) \tag{2.50}$$

It is useful to write this formula in the following way:

$$S(t, t_0|g) = T \exp(\int_{t_0}^{t} d\tau g h(\tau) \mathrm{V}(\tau)). \tag{2.51}$$

Let us suppose now that we are dealing with quantized classical theory. We assume that in classical theory the Hamiltonian $H(0)$ is positive and represented as a sum of a quadratic function of momenta p_i and a quadratic function of coordinates q^i; the interaction is represented by a polynomial of q^i.

Then we can write the corresponding quantum Hamiltonian in the form $\hat{H} = \hat{H}(0) + gh(t)\hat{V}$ where $\hat{H}(0) = \sum \epsilon_i a_i^* \hat{a}_i, \hat{V} = \sum \hat{V}_k,\ \hat{V}_k = \sum \gamma_{i_1,...i_k} : \hat{q}^{i_1} ... \hat{q}^{i_k} :$ (the indices run over a set X; this set can be finite or infinite, the coefficients γ are symmetric functions of indices). Here $\hat{a}^*$, $\hat{a}$ obey CCR for creation and annihilation operators, and $: A_1 ... A_n :$ stands for the normal product (we should express A_i in terms of $\hat{a}^*$, $\hat{a}$ and move all creation operators to the left assuming that all operators commute; for example, $: \hat{a}_i \hat{a}_j^* := \hat{a}_j^* \hat{a}_i$.)

Let us assume that $\hat{H}$ is a well-defined self-adjoint operator in Fock space.

We obtained expressions for $S(t, t_0|g)$ that involve chronological ordering of operators $\mathrm{V}(t) = e^{i\hat{H}(0)t} \hat{V} e^{-i\hat{H}(0)t}$, however, for calculations of matrix elements of $S(t, t_0|g)$ in Fock space it is convenient to write these expressions in terms of normal ordering. One can do this using the relation

$$T(\mathrm{q}^i(\tau)\mathrm{q}^j(\tau') =: \mathrm{q}^i(\tau)\mathrm{q}^j(\tau') : +G(i, \tau, j, \tau') \tag{2.52}$$

where $\mathrm{q}^i(t) = e^{i\hat{H}(0)t} \hat{q}^i e^{-i\hat{H}(0)t}$, and G is a function called propagator. (One should consider (2.52) as a definition of propagator. It is important to emphasize that G is a numerical function; this follows from the remark that $\mathrm{q}^i(\tau))$ is an element of Well algebra.) Applying (2.52) step by step we present (2.57) in normal form.

It is convenient to represent the answer graphically as a sum of diagrams. We depict V_k as a star (as a point (internal vertex) with k outgoing intervals; the ends of these intervals are called external vertices of the star). We connect in some way several external vertices of stars with external vertices of other stars by oriented intervals called edges. The resulting picture is called a diagram; the external vertices of stars that do not belong to any edge are called external vertices of the diagram. To every diagram, we assign an operator constructed in the following way. We assign to every external vertex of every star a pair (i, τ) where $i \in X, t_0 \leq \tau \leq t$, the times τ assigned to external vertices of the same star should coincide. To every star, we assign a function $h(\tau)\gamma_{i_1,...,i_k}$ where (i, τ) correspond to external vertices of the star. To an edge starting at the vertex (i', τ') and ending at vertex (i, τ) we assign the propagator $G(i, \tau, i', \tau')$. To an external vertex of a diagram we assign an operator $\mathrm{q}^i(\tau)$; we take the normal product of these operators.

Now we should multiply this normal product by functions corresponding to stars and by propagators corresponding to edges; we sum over indices $i \in X$ and integrate over $\tau \in [t_0, t]$. We obtain an operator presented in normal form; it corresponds to the diagram. To get $S_n(t, t_0)$ we sum contributions of all diagrams and divide by $n!$.

Notice that many diagrams give the same contributions (topologically equivalent diagrams), hence one can simplify the calculations summing over topological classes of diagrams.

Another simplification comes from remark that the contribution of disconnected diagram is equal to the product of the contributions of its connected components. We say that the sum of connected diagrams for $S(t, t_0|g)$ is a connected part of $S(t, t_0|g)$. It is easy to check that one gets the Wick symbol of $S(t, t_0|g)$ exponentiating the Wick symbol of its connected part.

2.8.2 Perturbation Theory for Green's Functions

Let us calculate the Green's functions of the time-independent Hamiltonian $\hat{H} = \hat{H}(0) + g\hat{V}$ in the framework of perturbation theory (we take $h(t) = 1$). We define Green's function as the expectation value of the chronological product $T(\hat{q}^{i_1}(t_1)...\hat{q}^{i_m}(t_m))$ in the ground state Φ of the Hamiltonian $\hat{H}$. (Here $\hat{q}^i(t) = e^{i\hat{H}t}\hat{q}^i e^{-i\hat{H}t}$ are Heisenberg operators.) To apply the above results we consider time-dependent Hamiltonian $\hat{H}_a = \hat{H}(0) + gh(at)\hat{V}$ where $h(t)$ is an even function obeying $h(t) = 1$ in a neghborhood of $t = 0$ and tending to zero as $t \to \infty$. Then for for small a we can use adiabatic perturbation theory; it follows that

$$\Phi = \lim_{a \to 0} P(a) S_a(0, -\infty|g)|0\rangle. \tag{2.53}$$

Here $P(a)$ is a phase factor, $|0\rangle$ is a ground state of $\hat{H}(0)$, and $S_a(t, t_0|g)$ stands for the evolution operator (in the interaction picture) corresponding to the Hamiltonian $\hat{H}_a$. (We keep notation $S(t, t_0|g)$ for the operator corresponding to $\hat{H}(0) + g\hat{V}$.)

We can say that the ground state Φ of the Hamiltonian $\hat{H}$ is obtained from the ground state $|0\rangle$ of the Hamiltonian $\hat{H}(0)$ using adiabatic dressing.

Let us introduce an adiabatic Green's function as the expectation value of chronological product

$$T\left(\mathrm{q}^{i_1}(t_1)...\mathrm{q}^{i_m}(t_m) \int_{-\infty}^{+\infty} d\tau \exp(gh(at)\mathrm{V}(\tau)\right)$$

in the ground state of $\hat{H}(0)$. Less precisely, but shorter this definition can be written in the following way

$$\langle 0|T\left(\mathrm{q}^{i_1}(t_1)...\mathrm{q}^{i_m}(t_m) S_a(+\infty, -\infty|g)\right)|0\rangle. \tag{2.54}$$

If the times $t_1, ..., t_m$ decrease then this expression can be presented in the form

$$\langle 0|S_a(+\infty, t_1|g)\mathrm{q}^{i_1}(t_1)S_a(t_1, t_2|g)\mathrm{q}^{i_2}(t_2)...\mathrm{q}^{i_m}(t_m)S_a(t_m, -\infty|g)|0\rangle.$$

For small a the operators $S_a(t_i, t_{i+1}|g)$ coincide with operators $S(t_i, t_{i+1}|g)$. (This follows from the assumption that $h(\tau)$ is equal to 1 for small τ.) Using the relation $\hat{q}^i(t) = S(0, t|g)\mathrm{q}^i(t)S(t, 0|g)$ we represent adiabatic Green's function (for small a and decreasing times) by the formula

$$\langle 0|S_a(+\infty, 0|g)\hat{q}^{i_1}(t_1)...\hat{q}^{i_m}(t_m)S_a(0, \infty|g)|0\rangle. \tag{2.55}$$

Applying (2.53), dividing (2.55) by the phase factor $\langle 0|S_a(+\infty, -\infty)|0\rangle$ and taking the limit $a \to 0$ we get Green's function.

We defined adiabatic Green's function in terms of chronological product. Using (2.52) we can express it as a constant term in a normal product. We obtain a perturbation series for adiabatic Green's function as a sum of diagrams. To divide by $\langle 0|S_a(+\infty, -\infty)|0\rangle$ one should exclude diagrams containing a connected component without external vertices. To get diagrams for Green's function we take the limit $a \to 0$; this means that we should replace $h(a\tau)$ by 1.

As a result, we obtain the following diagram technique for the calculation of Green's functions. An order n contribution to a Green's function depending on m times is a sum of diagrams with m points labeled by pairs $(i_1, t_1), ..., (i_m, t_m)$ where $i \in X, t \in \mathbb{R}$, n stars with external vertices labeled by pairs (i_r, τ) where $i_r \in X, \tau \in \mathbb{R}$ (τ is the same for all vertices of a star) and edges connecting two external vertices of different stars, external vertex of a star with one of m points or two of these m points (the diagrams with connected components without external vertices are excluded). To every star, we assign a factor $g\gamma_{i_1,...,i_k}$ where i_r are indices corresponding to external vertices of the star. To an edge starting at the vertex (i', τ') and ending at vertex (i, τ) we assign the propagator $G(i, \tau, i', \tau')$. We construct a function of $(i_1, t_1), ..., (i_m, t_m)$ taking a product of factors corresponding to stars and propagators corresponding to edges, summing over indices corresponding to external vertices of stars and integrating over times corresponding to stars. Green's function is a sum of functions corresponding to diagrams (a sum of diagrams, for the sake of brevity).

The diagram techniques can be generalized to the case when the index $i \in X$ is a continuous variable. Let us consider for example a classical Hamiltonian

$$\int \frac{1}{2}(\pi(\mathbf{x})^2 + (\nabla\phi(\mathbf{x}))^2)d\mathbf{x} + \sum_k g \int \gamma(\mathbf{x}_1, ..., \mathbf{x}_k)\phi(\mathbf{x}_1)...\phi(\mathbf{x}_k)d\mathbf{x}_1...\mathbf{x}_k. \tag{2.56}$$

Here $\phi(\mathbf{x})$ and $\pi(\mathbf{x})$ are scalar functions on $\mathbb{R}^d$ with Poisson brackets $\{\pi(\mathbf{x}), \phi(\mathbf{x}')\} = \delta(\mathbf{x} - \mathbf{x}')$, $\{\pi(\mathbf{x}), \pi(\mathbf{x}')\} = \{\phi(\mathbf{x}), \phi(\mathbf{x}')\} = 0$. It is not clear when the corresponding quantum Hamiltonian makes sense (even in non-relativistic case translation-invariant Hamiltonian considered in infinite volume does not specify an operator in Fock space). However, one can approximate this Hamiltonian by Hamiltonians of the type considered above, quantize them, calculate Green's functions, and define Green's functions of the quantized Hamiltonian (2.56) as a limit of Green's functions of approximate Hamiltonians. For these Green's functions, we obtain a diagram representation with the same diagrams; the only difference is that instead of summation over discrete indices $i \in X$ we have integration over continuous index $\mathbf{x} \in \mathbb{R}^d$. It is important to emphasize that there are some ambiguities in this procedure. Sometimes one can resolve these ambiguities by expressing the Green's functions not in terms of parameters appearing in the Hamiltonian (2.56) (bare coupling constants) but in terms of physical quantities.

2.8.3 Scattering Matrix from Adiabatic Scattering Matrix

The scattering matrix for translation-invariant formal Hamiltonian $\hat{H}(g) = \hat{H}(0) + g\hat{V}$ where $\hat{H}(0) = \int d\mathbf{p}\epsilon(\mathbf{p})a^*(\mathbf{p})a(\mathbf{p})$ can be expressed in terms of adiabatic scattering matrix $S_a^\Omega(g) = S_a^\Omega(+\infty, -\infty|g)$ of the Hamiltonian $\hat{H}^\Omega(g)$ obtained from $\hat{H}(g)$ by volume cutoff [17, 31]. (We can understand the volume cutoff, for example, as a replacement of integrals over $\mathbf{p}$ by summation over a cubic lattice $\mathbf{p} = \frac{\mathbf{n}}{L}$ where $\mathbf{n}$ is a vector having integer coordinates. When $L \to \infty$ the volume Ω tends to ∞.)

It is easy to give conditions when $\hat{H}^\Omega(g)$ specifies a self-adjoint operator in the Fock space $\mathcal{F}^\Omega$ (Fock representation of CCR for operators $a^*_\mathbf{p}$, $a_\mathbf{p}$ where $\mathbf{p}$ runs over the lattice.) We assume that these conditions are satisfied and the perturbation series for the adiabatic scattering matrix is well defined. In particular, this means that in the framework of perturbation theory the ground state $\Theta^\Omega(g)$ of $\hat{H}^\Omega(g)$ can be obtained by means of "adiabatic dressing" of the ground state θ of the Hamiltonian $\hat{H}(0)^\Omega$ and the stationary state $\phi_\Omega(\mathbf{p}, g)$ of $\hat{H}^\Omega(g)$ having minimal energy among states with momentum $\mathbf{p}$ can be obtained by means of "adiabatic dressing" of one-particle state $a^*_\mathbf{p}\theta$. The energy of the ground state $\Theta^\Omega(g)$ will be denoted by by $E_\Omega(g)$, the energy of $\phi_\Omega(\mathbf{p}, g)$ will be denoted by $E_\Omega(\mathbf{p}, g)$ and the difference $E_\Omega(\mathbf{p}, g) - E_\Omega(g)$ will be denoted $\epsilon_\Omega(\mathbf{p}|g)$.

Our definition specifies normalized vectors $\Theta_\Omega(g)$ and $\phi_\Omega(\mathbf{p}, g)$ only up to numerical factor; to get rid of this ambiguity we can impose the conditions

$$\langle\Theta_\Omega(g), \frac{\partial\Theta_\Omega(g)}{\partial g}\rangle = 0, \langle\phi_\Omega(\mathbf{p}, g); \frac{\partial\phi_\Omega(\mathbf{p}, g)}{\partial g}\rangle = 0.$$

Notice that the momentum operator $\mathbf{P}^\Omega = \sum \mathbf{p}a^*_\mathbf{p}a_\mathbf{p}$ has a discrete spectrum in finite volume. We use the notation $|\mathbf{p}_1, ..., \mathbf{p}_n\rangle$ for the standard basis in Fock space (recall that in finite volume $\mathbf{p}_i$ runs over a lattice). This basis consists of eigenvectors of the momentum operator.

Let us consider matrix elements

$$\langle\mathbf{p}_1...\mathbf{p}_m|S_{a,\Omega}|\mathbf{q}_1, ..., \mathbf{q}_n\rangle$$

of adiabatic scattering matrix in finite volume (in standard basis).

One can prove that these matrix elements multiplied by some factors tend to matrix elements of physical scattering matrix as $\Omega \to \infty$, $a \to 0$. (In finite volume matrix elements are defined on a lattice; we should extend a function on a lattice to a piecewise linear function on the whole space and take the limit in the sense of generalized functions.)

We introduce the following notations to describe the factors.

$$A_{a,\Omega} = \langle 0|\hat{S}_{a,\Omega}|0\rangle^{-\frac{1}{2}},\ B_{a,\Omega}(\mathbf{p}) = \langle \mathbf{p}|\hat{S}_{a\Omega}|\mathbf{p}\rangle^{\frac{1}{2}}$$

In these notations one can write down the expression for matrix elements of scattering matrix;

$$\begin{aligned} &\langle \mathbf{p}_1, ..., \mathbf{p}_m|\hat{S}|\mathbf{q}_1, .., \mathbf{q}_n\rangle \\ &= \lim_{a\to 0}\lim_{\Omega\to\infty} \frac{\langle \mathbf{p}_1, ..., \mathbf{p}_m|\hat{S}_{a,\Omega}|\mathbf{q}_1, .., \mathbf{q}_n\rangle A_{a,\Omega}^{2-m-n}}{B_{a,\Omega}(\mathbf{p}_1)...B_{a,\Omega}(\mathbf{p}_m)B_{a,\Omega}(\mathbf{q}_1)...B_{a,\Omega}(\mathbf{q}_n)} \end{aligned} \tag{2.57}$$

Notice that it follows from adiabatic perturbation theory that for $a \to 0$ we have

$$A_{a,\Omega} \approx e^{-i\int_{\infty}^{0} d\tau E_{\Omega}(h(\tau))},$$

$$B_{a,\Omega}(\mathbf{p}) \approx e^{i\int_{-\infty}^{0} d\tau (E_{\Omega}(\mathbf{p},h(\tau))-\epsilon(\mathbf{p}))},$$

the relation (2.57) remains correct if we replace the factors A, B by these approximations.

It is convenient to rewrite (2.57) in terms of connected scattering matrix $\hat{S}^{conn}$ and connected adiabatic scattering matrix $\hat{S}_{a,\Omega}^{conn}$. We represent these operators in normal form (take Wick symbol); corresponding coefficient functions will be denoted by $\nu_{m,n}$ and $\nu_{m,n}^{a,\Omega}$.

For $m + n > 2$ we obtain from (2.57) that

$$\begin{aligned} &\nu_{m,n}(\mathbf{p}_1, ..., \mathbf{p}_m, \mathbf{q}_1, ..., \mathbf{q}_n) \\ &= \lim_{a\to 0}\lim_{\Omega\to\infty} e^{i(r_{a,\Omega}(\mathbf{p}_1)+...+r_{a,\Omega}(\mathbf{p}_m)+r_{a,\Omega}(\mathbf{q}_1)+...+r_{a,\Omega}(\mathbf{q}_n))}\nu_{m,n}^{a,\Omega}(\mathbf{p}_1, ..., \mathbf{p}_m, \mathbf{q}_1, ..., \mathbf{q}_n) \end{aligned} \tag{2.58}$$

where

$$r_{a,\Omega}(\mathbf{k}) = \int_{-\infty}^{0} (\epsilon_{\Omega}(\mathbf{k}|h(\tau)) - \epsilon(\mathbf{k}))d\tau. \tag{2.59}$$

The formula (2.58) remains correct if we replace $r_{a,\Omega}(\mathbf{k})$ with $r_a(\mathbf{k}) = \lim_{\Omega\to\infty} r_{a,\Omega}(\mathbf{k})$. One can give an independent definition of $r_a(\mathbf{k})$ by the formula

$$r_a(\mathbf{k}) = \int_{-\infty}^{0} (\epsilon(\mathbf{k}|h(\tau)) - \epsilon(\mathbf{k}))d\tau. \tag{2.60}$$

where $\epsilon(\mathbf{k}|h)$ stands for one-particle energy (dispersion law) for the Hamiltonian $\hat{H}(0) + gh\hat{V}$.

The proof of (2.58) is based on the representation of connected scattering matrices as sums of tree diagrams having 1PI diagrams as vertices and two-point Green's functions as propagators. (By definition, 1PI diagram remains connected after cutting one internal edge.) A crucial fact used in the proof is a remark that in $(\mathbf{p}, t)$-representation 1PI diagram is small it at least for one pair of arguments t_i, t_j the difference $t_i - t_j$ is large. (See [31] for details.)

2.8.4 Perturbation Theory for Generalized Green's Functions

In Sect. 2.8.2 we worked in Hilbert space. However, studying generalized Green's functions we need the algebraic approach. It allows us to work with translation-invariant Hamiltonians directly in infinite volume.

We will use the notion of adiabatic generalized Green's function introduced below.

In the definition of generalized Green's functions, we assumed that spatial translations and time translations act as automorphisms of the algebra $\mathcal{A}$, hence they act also as linear transformations of the space $\mathcal{L}$ of continuous linear functionals on $\mathcal{A}$. We say that infinitesimal temporal translation in $\mathcal{L}$ is a "Hamiltonian" and denote it by H; this means that the temporal translation (evolution operator) has the form $\exp(H\tau)$. Similarly, we can say that infinitesimal spatial translation in $\mathcal{L}$ is "momentum"; we denote it by $\mathbf{P}$.

Let us assume that these data depend on a parameter g. For simplicity, we suppose that spatial translations do not depend on g, and the "Hamiltonian" has the form $H(g) = H(0) + gV$. We define the evolution operator $U_a(t, t_0|g)$ and the evolution operator in the interaction picture $S_a(t, t_0|g)$ as evolution operators corresponding to time-dependent "Hamiltonian" $H(0) + gh(a\tau)V$. (Here again $h((\tau)$ is equal to 1 in a neighborhood of $\tau = 0$ and tends to zero as $\tau \to \infty$.) It is easy to verify that the expression

$$\omega(g) = \lim_{a\to 0} U_a(0, -\infty|g)\omega_0 = \lim_{a\to 0} S_a(0, -\infty|g)\omega_0 \tag{2.61}$$

where ω_0 is a translation-invariant stationary state for the "Hamiltonian" $H(0)$ is a translation-invariant stationary state for the "Hamiltonian" $H(0) + gV$. (We assume that the limit in (2.61) exists.) We say that $\omega(g)$ is obtained from ω_0 using adiabatic dressing.

We define adiabatic generalized Green's function by the formula

$$\langle 1|T\Big(\tilde{\mathrm{A}}_1(\mathbf{x}_1, t_1)\ldots\tilde{\mathrm{A}}_m(\mathbf{x}_m, t_m)\mathrm{B}_1(\mathbf{x}'_1, t'_1)\ldots\mathrm{B}_n(\mathbf{x}'_n, t_n)S_a(+\infty, -\infty|g)\Big)|\omega_0\rangle \tag{2.62}$$

If a is small and $t_i > t_{i+1}, t'_j > t'_{j+1}$ then (2.62) can be written in the form

$$\langle 1|S_a(+\infty,0|g)\tilde{A}_1(\mathbf{x}_1,t_1|g)...\tilde{A}_m(\mathbf{x}_m,t_m|g)B_1(\mathbf{x}'_1,t'_1|g)...B_n(\mathbf{x}'_n,t_n|g)S_a(0,-\infty|g)|\omega_0\rangle \tag{2.63}$$

(We used the relations $\mathrm{A}(t)=e^{-H(0)t}Ae^{H(0)t}$, $A(t|g)=e^{-H(g)t}Ae^{H(g)t}=S(0,t|g)\mathrm{A}(t)S(t,0|g)$, $S_a(t,t_0|g)=S(t,t_0|g)$ for small a.

It follows from (2.63), (2.61) and the relation

$$\langle 1|S_a(+\infty,0|g)=\langle 1| \tag{2.64}$$

that *adiabatic generalized Green's functions tend to generalized Green's functions as* $a\to 0$. (The relation (2.64) follows from the remark that the unit element of algebra is invariant with respect to any evolution operator because it is invariant under any automorphism.)

Similar results can be proved for GGreen functions defined in Sect. 2.5.2. In addition to the state ω_0 that is translation-invariant and stationary with respect to $H(0)$ we fix a translation-invariant and stationary element α_0 of the algebra $\mathcal{A}$. Using α_0 and ω_0 we define adiabatic GGreen functions

$$\langle\alpha_0|T\Big(\tilde{\mathrm{A}}_1(\mathbf{x}_1,t_1)...\tilde{\mathrm{A}}_m(\mathbf{x}_m,t_m)\mathrm{B}_1(\mathbf{x}'_1,t'_1)...\mathrm{B}_n(\mathbf{x}'_n,t_n)S_a(+\infty,-\infty|g)\Big)|\omega_0\rangle \tag{2.65}$$

If a is small and $t_i>t_{i+1}, t'_j>t'_{j+1}$ then (2.65) can be written as follows

$$\langle\alpha_0|S_a(+\infty,0|g)\tilde{A}_1(\mathbf{x}_1,t_1|g)...\tilde{A}_m(\mathbf{x}_m,t_m|g)B_1(\mathbf{x}'_1,t'_1|g)...B_n(\mathbf{x}'_n,t_n|g)S_a(0,-\infty|g)|\omega_0\rangle \tag{2.66}$$

Assuming that there exists a limit

$$\langle\alpha(g)|=\lim_{a\to 0}\langle\alpha_0|S_a(+\infty,0|g)$$

we obtain that the adiabatic GGreen function tends to GGreen function corresponding to $\alpha(g)$, $\omega(g)$ as $a\to 0$.

Chapter 3
Poincaré Group. Relativistic Theories

This chapter is a short review of relativistic local theories in algebraic approach.

3.1 Introduction

In the present chapter, we consider relativistic (= Poincaré-invariant) theories. Standard textbooks in quantum field theories deal only with such theories; moreover, it is assumed in these books that the theories under consideration are local. In the algebraic approach, we assume that the Poincaré group is a subgroup of the automorphism group of algebra $\mathcal{A}$. The notion of field is not necessary, locality is replaced by asymptotic commutativity or by cluster property. (Notice that the condition of strong asymptotic commutativity imposed in Chap. 2 is violated in theories with massless particles.) Poincaré group contains the group of spatial and temporal translations; this allows us to apply the theory of scattering of elementary excitations of translation-invariant stationary state ω developed in Chap. 2. We will impose a stronger condition that ω is Poincaré-invariant; then the Poincaré group acts by unitary transformations in Hilbert space $\mathcal{H}$ obtained by GNS construction.

Let us sketch the proof that the corresponding Møller matrices and scattering matrix are Poincaré-invariant. It is sufficient to prove that the in-states do not change when we change the inertial system of coordinates. Recall that an in-state is defined as a limit of vectors (2.20) as $t \to -\infty$. To obtain the vector (2.20) we apply several operators $B(f_i, t)$ to the vector $\theta \in \mathcal{H}$ corresponding to ω. Let us denote by $B'(f_i', t')$ similar operators in another inertial system obeying

$$B(f_i, t)\theta = B(f_i', t')\theta. \tag{3.1}$$

A. Schwarz, *Quantum Mechanics and Quantum Field Theory in Algebraic and Geometric Approaches*, SpringerBriefs in Physics,
https://doi.org/10.1007/978-3-031-67915-5_3

(We assume that $B\theta$ and $B'\theta$ are one-particle states, then $B(f_i, t)\theta$ and $B'(f'_i, t')\theta$ also are one-particle states that do not depend on t and t'.) Now we can prove that replacing in (2.20) some operators $B(f_i, t)$ by operators $B'(f'_i, t')$ we obtain the same limit as $t, t' \to -\infty$. The proof is based on the remark that due to asymptotic commutativity operators $B(f_i, t)$ and $B(f'_j, t')$ commute in the limit $t, t' \to -\infty$.(We assume the functions f_i, f'_j do not overlap.) If we replace the operator $B(f_i, t)$ by the operator $B'(f'_i, t')$ in the rightmost place nothing changes due to the assumption (3.1). To give the proof in the general case we move the operators to the right and use (3.1) again.

Very similar considerations can be applied in the case when the algebra $\mathcal{A}$ is asymptotically anti-commutative.

Notice, that for local theories the algebraic approach can be based on Araki-Haag-Kastler axioms. In these axioms one assigns an algebra to every bounded domain in Minkowski space. The union of these algebras is strongly asymptotically commutative if masses in the theory are bounded below by a positive number (there exists a mass gap).

3.2 Lorentz Group. Poincaré Group

By definition, Lorentz group L is a connected component of the group $O(1, 3)$, i.e. of the group of linear transformations of $\mathbf{R}^4$ that preserve a quadratic form

$$h_{ab}x^a x^b = (x^0)^2 - (x^1)^2 - (x^2)^2 - (x^3)^2. \tag{3.2}$$

(Here $x^0 = t$ plays the role of time and $(x^1, x^2, x^3) = (x, y, z)$ are spatial coordinates.)

The tensors h_{ab} and h^{ab} represented by diagonal matrices with diagonal entries $1, -1, -1, -1$ are Lorentz-invariant; we can use them to lower and raise indices of tensors.

The Poincaré group $\mathcal{P}$ is a group of transformations of $\mathbf{R}^4$ having the form $x' = \Lambda x + a$ where $\Lambda \in L$, $a \in \mathbf{R}^4$. This is a connected component of the group $\hat{\mathcal{P}}$ consisting of affine transformations preserving space-time interval $ds^2 = h_{ab}dx^a dx^b$.

More generally, one can define Lorentz group as a connected component of the group $O(1, d)$ of linear transformations of the space $\mathbb{R}^{1+d}$ that preserve a quadratic form $(x^0)^2 - (x^1)^2 - ... - (x^d)^2$. Similarly one can generalize the notion of Poincaré group.

The group L is closely related to the group $SL(2, \mathbb{C})$. To describe this relation we consider a representation of the group $SL(2, \mathbb{C})$ in the space $\mathbf{H}$ of Hermitian 2×2 matrices:

$$\varphi_A(X) = AXA^*. \tag{3.3}$$

where $A \in SL(2,\mathbb{C})$, $X \in \mathbf{H}$. A matrix $X \in \mathbf{H}$ has a unique representation in the form

$$X = \begin{pmatrix} x^0 + x^3 & x^1 - ix^2 \\ x^1 + ix^2 & x^0 - x^3 \end{pmatrix} \tag{3.4}$$

hence the space $\mathbf{H}$ is a real four-dimensional vector space with coordinates (x^0, x^1, x^2, x^3).

If $\widetilde{X} = AXA^*$ and $A \in SL(2,\mathbb{C})$ then $\det \widetilde{X} = \det A \cdot \det X \cdot \det A^* = \det X$. This means that the transformation $\varphi_A : \mathbf{H} \to \mathbf{H}$ preserves the quadratic form $\det X = (x^0)^2 - (x^1)^2 - (x^2)^2 - (x^3)^2$.

In other words, φ can be considered as a homomorphism $\varphi : SL(2,\mathbb{C}) \to O(1,3)$. It is easy to verify that $\varphi(SL(2,\mathbb{C})) = L$, hence L is isomorphic to $SL(2,\mathbb{C})/Ker\varphi$. where $Ker\varphi$ consists of two matrices: $\mathrm{Ker}\ \varphi = \{1, -1\}$.The isomorphism $L = SL(2,\mathbb{C})/\{1,-1\}$ allows us to identify representations of $SL(2,\mathbb{C})$ with single-valued and two-valued representations of L. In what follows talking about representations of the Lorentz group, Poincaré group, and orthogonal groups we have in mind single-valued and two-valued representations.

The group $SL(2,\mathbb{C})$ has two non-equivalent two-dimensional representations: a matrix $A \in SL(2,\mathbb{C})$ can be represented either by the same matrix A or by complex conjugate matrix $\bar{A}$ (vector representation and its complex conjugate). The elements of corresponding representation spaces will be denoted by η^{α} and $\eta^{\dot{\alpha}}$. All other irreducible representations can be obtained as tensor products of p-th symmetric power of vector representation and q-th symmetric power of its complex conjugate. The elements of corresponding representation spaces are tensors $\phi^{\alpha_1,\dots,\alpha_p,\dot{\beta}_1,\dots,\dot{\beta}_q}$ that are symmetric with respect to α_1, ..., α_p and with respect to $\dot{\beta}_1$, ..., $\dot{\beta}_q$. Corresponding representations of Lorentz group are single-valued for even $p+q$ and two-valued for odd $p+q$. The representation of $SL(2,\mathbb{C})$ in the space $\mathbf{H}$ of Hermitian matrices is an irreducible representation with $p = q = 1$, hence Lorentz vectors can be identified with tensors $\phi^{\alpha,\dot{\beta}}$.

Notice that tensors $\epsilon_{\alpha,\beta}$, $\epsilon^{\alpha,\beta}$, $\epsilon_{\dot{\alpha},\dot{\beta}}$, $\epsilon^{\dot{\alpha},\dot{\beta}}$ represented by antisymmetric matrices are $SL(2,\mathbb{C})$-invariant. One can use them to lower and raise indices establishing equivalence of vector and covector representations.

3.3 Particles in Relativistic Theories

Let us consider a relativistic theory in the algebraic approach. This means that we consider a $*$-algebra $\mathcal{A}$ and an embedding of Poincaré group $\mathcal{P}$ into the group of automorphisms of $\mathcal{A}$. Let us fix a $\mathcal{P}$-invariant state (normalized positive functional ω on $\mathcal{A}$); the Hilbert space $\bar{\mathcal{H}}$ obtained by GNS-construction contains vector θ corresponding to the state ω. The Poincaré group descends on $\bar{\mathcal{H}}$ as a group of unitary operators. The translation group is a subgroup of $\mathcal{P}$; therefore we obtain spatial translations $T_{\mathbf{a}}$ and time translations T_τ as unitary operators in $\bar{\mathcal{H}}$. We define momentum operator $\mathbf{P}$ and energy operator H by the formulas $T_{\mathbf{a}} = \exp(i\hat{\mathbf{P}}\mathbf{a})$, $T_\tau = \exp(-i\hat{H}\tau)$.

By general definition an elementary excitation of ω is a generalized vector function $\Phi(\mathbf{p})$ taking values in $\mathcal{H}$ and obeying $\hat{\mathbf{P}}\Phi(\mathbf{p}) = \mathbf{p}\Phi(\mathbf{p})$, $\hat{H}\Phi(\mathbf{p}) = \epsilon(\mathbf{p})\Phi(\mathbf{p})$. We assume that ω is a ground state (i.e. $\hat{H}$ is positive definite), therefore we identify elementary excitations with particles. It is easy to check that an element of Poincaré group transforms a particle into a particle, hence one-particle subspace $\mathcal{H}_1$ (the subspace spanned by vectors $\Phi(\phi) = \int d\mathbf{p}\phi(\mathbf{p})\Phi(\mathbf{p})$ where Φ runs over particles and ϕ runs over test functions) is Poincaré -invariant. If $\mathcal{H}_1$ is a space of irreducible representation of $\mathcal{P}$ we say that it specifies a relativistic particle; if this space is a direct sum of irreducible representations we are dealing with several relativistic particles. (The identification of relativistic particles with irreducible representations of $\mathcal{P}$ suggested by Wigner [39] does not always agree with standard terminology. In particular, photon corresponds to a reducible representation of the group $\mathcal{P}$, however, it corresponds to an irreducible representation of the extended Poincaré group $\hat{\mathcal{P}}$.) The operator $\hat{H}^2 - \hat{\mathbf{P}}^2$ commutes with operators corresponding to elements of Poincaré group and its Lie algebra hence it is constant on every irreducible representation. This means that for all elementary excitations inside of irreducible representation, we have $\epsilon(\mathbf{p})^2 - \mathbf{p}^2 = m^2$ where m is a constant (mass of the particle). It follows from the assumption that ω is a ground state that m is a non-negative real number and $\epsilon(\mathbf{p}) = +\sqrt{\mathbf{p}^2 + m^2}$. Let us assume that a representation of $\mathcal{P}$ is spanned by a finite number r of elementary excitations $\Phi_1(\mathbf{p}), \ldots, \Phi_r(\mathbf{p})$. This means that every point in the space of this representation can be uniquely represented in the form

$$\int d\mathbf{p} c_1(\mathbf{p})\Phi_1(\mathbf{p}) + \ldots + \int d\mathbf{p} c_r(\mathbf{p})\Phi_r(\mathbf{p}). \tag{3.5}$$

We say that unitary representations of $\mathcal{P}$ obeying the conditions above (positive definiteness of $\hat{H}$ and finiteness of r) are admissible.

It is convenient to change notations assuming that Φ_i is defined on on the subspace O_m of Minkowski space $\mathbb{R}^4$ consisting of of points of the form $p = (+\sqrt{\mathbf{p}^2 + m^2}, \mathbf{p}) = (p_0, \mathbf{p})$. For $m > 0$ this is an orbit of the point $(m, 0, 0, 0)$ with respect to the action of the Lorentz group, for $m = 0$ this is an orbit of the point $(1, 0, 0, 1)$. The stabilizer of Lorentz group at the point $(m, 0, 0, 0)$ ("litle group" in the terminology of physicists) is isomorphic to the group of spatial rotations $SO(3)$. The stabilizer at the point $(1, 0, 0, 1)$ is isomorphic to the group $ISO(2)$ of rigid motions of a plane. We will be interested only in the compact part of the stabilizer; it is isomorphic to $SO(2)$; the name "little group" is reserved for $SO(2)$. In new notations (3.5) can be written as

$$\int d\mu c_1(p)\Phi_1(p) + \ldots + \int d\mu c_r(p)\Phi_r(p) = \int d\mu \langle c(p), \Phi(p)\rangle. \tag{3.6}$$

In this formula p runs over O_m and $d\mu$ stands for Lorentz-invariant volume element $d\mathbf{p}/\sqrt{2p_0}$ on O_m. We consider $(c_1(p), \ldots, c_(p))$ and $(\Phi_(p), \ldots, \Phi_r(p)$ as r-dimensional vectors $c(p)$ and $\Phi(p)$. The action of space and time translation $T_\tau T_\mathbf{a}$ on $c(p)$ can be written as multiplication by $\exp(-i\tau p_0 + i\mathbf{a}\mathbf{p})$. It follows from this fact the a Lorentz transformation Λ maps $c(p)$ into $B(\Lambda, p)c(\Lambda p)$ where $B(\Lambda, p))$ is an $r \times r$ matrix. Matrices $B(\Lambda, p)$ where Λ belongs to the stabilizer $Stab(p)$ of Lorentz group at the point p specify an r-dimensional unitary representation of the stabilizer; it follows from irreducibility of the representation of $\mathcal{P}$ that the representation of the stabilizer is also irreducible. One can prove that an irreducible unitary finite-dimensional representation of the stabilizer specifies an irreducible unitary representation of the Poincaré group up to equivalence [39]. More precisely, an equivalence class of admissible representations of $\mathcal{P}$ with the mass $m > 0$ is specified by the dimension r of irreducible representation of the "little group" $SO(3)$. (Recall, that odd-dimensional representations of $SO(3)$ are single-valued and even-dimensional representations are two-valued). The corresponding particle has mass m and spin $s = (r - 1)/2$. For $m = 0$ the equivalence classes of admissible representations of $\mathcal{P}$ correspond to irreducible single-valued and two-valued representations of the "little group" $SO(2)$. All these representations of $SO(2)$ are one-dimensional, they are labeled by an integer k; if an element of $SO(2)$ corresponds to the rotation angle α then it is represented by $e^{ik\alpha}$. For even k such a representation is single-valued, for odd k it is two-valued. (Notice that by removing the restriction that r is finite we obtain a new class of representations, continuous spin representations. These representations correspond to the representations of the stabilizer $ISO(2)$ that cannot be obtained by combining the homomorphism $ISO(2) \to SO(2)$ and a representation of $SO(2)$. Known particles do not correspond to continuous spin representations.)

Defining the little group $SO(2)$ we considered the stabilizer of the vector $(1, 0, 0, 1)$. With this choice, the little group can identified with rotations around z-axis and the number $k/2$ with z-component of the spin. In general, $k/2$ is equal to helicity (to the projection of spin in the direction of momentum).

3.4 Examples

Let us consider a space of complex functions defined on Minkowski space. The action of the Poincaré group on the Minkowski space induces a linear representation of this group in the space of functions. More generally, instead of complex functions, we can consider functions taking values in a representation of the Lorentz group; again we obtain a representation of the Poincaré group in the space of functions. The representations of this kind are reducible; we can find a subrepresentation imposing the condition that functions obey some Poincaré-invariant linear equations (such equations come from Poincaré-invariant quadratic action functionals).

An equation where both the left-hand side and right-hand side transform according to the same representation of the Lorentz group is Lorentz-invariant. If it is also translation-invariant (the coefficients are constant) then it is Poincaré-invariant.

Let us give some examples.

3.4.1 Scalar Field

Let us consider Poincaré -invariant equation

$$(\Box + m^2)\phi(x) = 0$$

(Klein-Gordon equation). Here $\phi(x) = \phi(x^0, \mathbf{x})$ stands for complex-valued function on Minkowski space and

$$\Box = h^{ab}\frac{\partial^2}{\partial x^a \partial x^b} = \frac{\partial^2}{\partial t^2} - \nabla^2$$

denotes the d'Alembert operator.

The space of solutions to the Klein-Gordon equation is a representation space for a representation of the Poincaré group. (Notice that our description is ambiguous because we did not specify the class of functions we consider.)

In momentum representation (= after Fourier transform) the Klein-Gordon equation takes the form $(p^2 - m^2)\phi(p) = 0$ where $p^2 = (p_0)^2 - \mathbf{p}^2$. The solution of this equation is a generalized function that vanishes in the complement of the set $\mathcal{K}_m$ singled out by the equation $p^2 = m^2$. It can be represented as a function on $\mathcal{K}_m$ multiplied by $\delta(p^2 - m^2)$. (For $m > 0$ the set $\mathcal{K}_m$ is a hyperboloid consisting of two components. For $m = 0$ this set is a union of two cones.) To construct an irreducible unitary representation for $m > 0$ we consider square-integrable functions on the upper half of hyperboloid ($p_0 = +\sqrt{\mathbf{p}^2 + m^2} > 0$) (the integration is defined with respect to Lorentz-invariant volume element $\frac{d\mathbf{p}}{\sqrt{2p_0}}$). The Hilbert space KG_+ of square-integrable functions on the upper half of the hyperboloid is a representation space of irreducible admissible representation corresponding to the trivial representation of little group $SO(3)$. Similar construction works for $m = 0$; it leads to irreducible unitary representation corresponding to the trivial representation of the little group $SO(2)$ (helicity is equal to zero).

3.4.2 Particles with Spin $\frac{1}{2}$

The equations

$$\partial_{\alpha,\dot{\beta}}\eta^{\alpha} = 0, \ \partial_{\alpha,\dot{\beta}}\eta^{\dot{\beta}} = 0 \tag{3.7}$$

describe massless neutrino and antineutrino.

Here $\eta^{\alpha}(x)$, $\eta^{\dot{\beta}}(x)$ are functions on Minkowski space taking values in two-dimensional two-valued representations of Lorentz group, $\partial_{\alpha,\dot{\beta}}$ can be identified with partial derivative $\partial_a = \partial/\partial x^a$. The Eq. (3.7) are Poincaré-invariant, hence the corresponding spaces of solutions are representation spaces of representations of Poincaré group. Introducing an appropriate scalar product we obtain two complex-conjugate unitary representations of Poincaré group; these representations correspond to representations of the little group $SO(2)$ with $k = \pm 1$ (with helicity $\frac{1}{2}$ for neutrino and $-\frac{1}{2}$ for antineutrino).

To describe massive particles having spin $\frac{1}{2}$ one can consider a pair of functions $\eta^{\alpha}(x), \eta^{\dot{\beta}}(x)$ and equations

$$\partial_{\alpha,\dot{\beta}}\eta^{\alpha} = \text{const} \cdot \eta_{\dot{\beta}}, \partial_{\alpha,\dot{\beta}}\eta^{\dot{\beta}} = \text{const} \cdot \eta_{\alpha}. \tag{3.8}$$

However, it is more convenient to write these equations as Dirac equation

$$\gamma^{a}\frac{\partial\psi}{\partial x^{a}} = m\psi \tag{3.9}$$

where $\psi(x)$ is a function taking values in spinor representation of the group $SO(1,3)$ described in Sect. 1.12 and γ^a are four-dimensional Dirac matrices obeying the relations of Clifford algebra: $[\gamma^a, \gamma^b]_+ = 2h^{ab}$. (Recall that by definition γ^a transforms as a vector with respect to orthogonal transformations, hence the Dirac operator $D = \gamma^a\partial_a$ is a scalar; these means that both summands in (3.9) transform according spinor representation). The square of the Dirac operator is the d'Alember operator $\Box$, therefore every solution of the Dirac equation (3.9) is a solution of the Klein-Gordon equation and can be considered as a four-component function on the hyperboloid $p^2 = m^2$. For an appropriate choice of scalar product solutions of (3.9) constitute the representation space of reducible unitary representation of $\mathcal{P}$. This representation is a direct sum of representations Dir_+ and Dir_- corresponding to the upper and lower halves of the hyperboloid. The representation Dir_+ is an admissible representation corresponding to the two-dimensional (spinor) representation of the little group $SO(3)$. (It describes a massive particle of mass m.)

The considerations of Sect. 1.12 allow us to define spinor representations of the Lorentz group in any dimension. Using this remark we can write down the Eq. (3.9) (Dirac equation) in any dimension.

Notice that for the Dirac field $\psi(x)$ we can define divergence-free current $j^a(x) = e\bar{\psi}(x)\gamma^a\psi(x)$ where $\bar{\psi}(x) = \psi(x)^*\gamma^0$. This can be checked directly or derived from Noether's theorem. (The Dirac equation and the corresponding action functional are invariant with respect to the transformation $\psi(x) \to e^{i\alpha}\psi(x)$. Noether's theorem gives a construction of divergence-free current in this situation; it can be applied also to complex scalar field.)

3.4.3 Electromagnetic Field

To write Maxwell equations in Lorentz-invariant form we introduce rank 2 antisymmetric tensor F_{ab} with components expressed in terms of electric field $\mathbf{E}$ and magnetic field $\mathbf{H}$; namely $F_{01} = E_1$, $F_{02} = E_2$, $F_{03} = E_3$, $F_{12} = -H_3$, $F_{13} = H_2$, $F_{23} = -H_1$. Maxwell equations for F_{ab} can be written as follows:

$$\partial_a F_{bc} + \partial_b F_{ca} + \partial_c F_{ab} = 0, \tag{3.10}$$

$$\partial_a F^{ab} = j^b, \tag{3.11}$$

where j is the four-current obeying $\partial_a j^a = 0$. It follows from (3.10) that F_{ab} can be represented in the form $F_{ab} = \partial_a A_b - \partial_b A_a$ where the 4-vector A_a is called electromagnetic potential. The potential is defined only up to gauge transformations $A_a \to A_a + \partial_a \epsilon$ where $\epsilon(x)$ is an arbitrary function. One can use the freedom in the choice of potential (gauge freedom) to simplify Maxwell equations. If we impose a Lorentz- invariant condition

$$\partial^a A_a = 0 \tag{3.12}$$

(Lorenz gauge) we obtain the simplest form of Maxwell equation $\square A_a = j_a$.

Let us consider free electromagnetic field ($j^a = 0$). Then the space of solutions to the equation

$$\square A_a = 0 \tag{3.13}$$

obeying Lorenz gauge condition (3.12) can be regarded as a representation space of a representation of Poincaré group. After Fourier transformation, these solutions can be written as products of functions $A^a(p)$ on the cone $p_0^2 = \mathbf{p}^2$ by delta-function $\delta(p_0^2 - \mathbf{p}^2)$; we consider a representation of $\mathcal{P}$ on the space of functions $A^a(p)$ on the upper half of the cone ($p_0 = +|\mathbf{p}|$) obeying $p^a A_a(p) = 0$. Let us define a scalar product in this space by the formula $\langle A, B\rangle = -\int h_{ab} A^a(p) \bar{B}^b(p) dp$ where dp stands for Lorentz-invariant volume element on the cone. This scalar product is non-negative, but degenerate; the functions of the form $p^a \epsilon(p)$ are null vectors of the scalar product (they are gauge equivalent to zero). Factorizing square-integrable functions with respect to null vectors we obtain a unitary representation of $\mathcal{P}$ that corresponds to the photon. This unitary representation is a direct sum of two admissible irreducible representations (with helicities $+1$ and -1). To check this we notice that the little group $SO(2)$ acts on the coefficients $(A_0(p), A_1(p), A_2(p), A_3(p))$ where $p = (1, 0, 0, 1)$ as a rotation in the plane A_1, A_2. (It follows from Lorenz condition that $A_)(p) + A_3(p) = 0$. Factorizing with respect to null vectors having the form $(\epsilon, 0, 0, \epsilon)$ we see that one can assume that $A_0(P) = A_3(p) = 0$.) Over complex numbers the two-dimensional representation of $SO(2)$ can be represented as a direct sum of one-dimensional representations.)

3.5 Free Theories

Recall that for every pre-Hilbert space $\mathcal{E}$ we can define Weyl algebra and Clifford algebra as $*$-algebras with generators $a(f), a^*(g)$ depending linearly of $f, g \in \mathcal{E}$ and obeying canonical (anti)commutation relations (CCR or CAR)

$$[a(f, a(g)]_\mp = [a^*(f), a^*(g)]_\mp = 0, [a(f), a^*(g)]_\mp = (\bar{f}, g).$$

We can consider Fock representation of Weyl/Clifford algebra $\mathcal{A}$ (in other words, we can quantize $\mathcal{E}$ using CCR or CAR). Notice, that corresponding operators $\hat{a}(f), \hat{a}^*(g)$ are bounded for Clifford algebra and unbounded for Weyl algebra, however, considering the exponential form of Weyl algebra (Sect. 1.10) we also can work with bounded operators.

Let us denote by $\mathcal{E}$ the representation space of an admissible unitary representation of the Poincaré group. We can use the space $\mathcal{E}$ (or a dense subset of this space) to construct Weyl or Clifford algebra and corresponding Fock space $\mathcal{F}$. Then the Fock space can be regarded as a representation space of representation of Poincaré group (this follows immediately from the remark that $\mathcal{F}$ can be considered as a direct sum of symmetric or antisymmetric tensor powers of $\mathcal{E}$). The Fock vacuum θ is Poincaré-invariant and vectors of the form $a^*(f)\theta$ are one-particle states. Working with Weyl algebras we obtain relativistic bosons, in the case of Clifford algebras, we get relativistic fermions. Notice that in relativistic local field theory, we should use Clifford algebras if the representation of $\mathcal{P}$ is two-valued and Weyl algebra if this representation is single-valued. (If the representation is a direct sum of single-valued and two-valued representations we should use a tensor product of Weyl and Clifford algebras.) In our setting (as well as in non-relativistic quantum mechanics) this statement (spin-statistics theorem) does not apply.

We will use the space $\mathcal{S} \subset \mathcal{E}$ of smooth fast decreasing functions to construct the algebra $\mathcal{A}$. Recall that $\mathcal{E}$ can be realized as a space of vector-valued functions on $\mathbb{R}^3$, hence we can consider $\hat{a}(f)$ and $\hat{a}^*(f)$ as vector-valued generalized functions: $\hat{a}(f) = \sum_k d\mathbf{p} f^k \hat{a}_k(\mathbf{p}), \hat{a}^*(g) = \sum_k d\mathbf{p} g^k \hat{a}_k(\mathbf{p})$. Momentum and energy operators can be written in the form $\mathbf{P} = \sum_k \int \mathbf{p} a_k^*(\mathbf{p}) a_k(\mathbf{p}) d\mathbf{p}$, $H = \sum_k \int \epsilon_(\mathbf{p}) a_k^*(\mathbf{p}) a_k(\mathbf{p}) d\mathbf{p}$, where $\epsilon_k(\mathbf{p}) = \sqrt{\mathbf{p}^2 + m_k^2}$. This means that we are dealing with free theory.

Let us assume that the representation of $\mathcal{P}$ in $\mathcal{E}$ is a direct sum of a finite number of irreducible representations corresponding to massive particles. We will prove that in this case the algebra $\hat{\mathcal{A}}\theta)$ of operators representing $\mathcal{A}$ in the Fock space is strongly asymptotically (anti)commutative.

Let us start with the case of when we quantize $\mathcal{E}$ using CCR. In the representation of CCR in Fock space we consider an algebra generated by unitary operators $V_f = \exp(-\hat{a}(\bar{f}) + \hat{a}^*(f))$. It is easy to check that $V_f V_g = V_g V_f \exp(-\langle g, f\rangle + \langle f, g\rangle)$. To prove that operators V_f and V_g strongly asymptotically commute we should shift one of these operators in space by $\mathbf{a}$ and in time by t and prove that the norm of the commutator is less than $C_n(t)||\mathbf{a}||^{-n}$ where n is any number and $C_n(t)$ is a polynomial. This follows from the remark that the same estimate holds for $\int d\mathbf{p} e^{i\mathbf{pa}-ip_0t)}\rho(\mathbf{p})$ where $p_0 = \sqrt{\mathbf{p}^2 + m^2}$ and $\rho = f\bar{g}(2p_0)^{-1/2}$. (To get this estimate we integrate by parts n times.)

It follows that the algebra generated by V_f is strongly asymptotically commutative (asymptotic commutativity of an algebra follows from asymptotic commutativity of generators).

Notice that in the of massless particles the function $\rho(\mathbf{p})$ is not smooth at the point $\mathbf{p} = 0$, hence the above proof fails, however, some weaker statements about asymptotic commutativity remain correct.

If we quantize $\mathcal{E}$ using CAR the same arguments can be used to show that the Clifford algebra generated by $a(f), a^*(g)$ where $f, g \in \mathcal{S}$ is strongly asymptotically anticommutative.

The scattering matrix in free theory is trivial. To verify this one can notice, for example, that $B(f, t)$ does not depend on time if we take $B = \hat{a}^*(g)$. Another way to check this fact is to use the remark that truncated correlation functions for free theories vanish.

3.6 Interacting Field Theories

We will give a very short review of interacting relativistic field theories taking as an example non-linear scalar field $\phi(x)$ on four-dimensional space-time.

The simplest Poincaré invariant non-linear classical equation can be written in the form

$$\Box\phi + m^2\phi + \sum_{k=3}^{k=s} \lambda_k \phi^{k-1} = 0 \tag{3.14}$$

It comes from the action functional $S(\phi) = \int d^4x(\frac{1}{2}h^{ab}\partial_a\partial_b - m^2\phi^2 - P(\phi)$ where $P(\phi) = \sum_{k=3}^{k=s} \frac{1}{k!}\lambda_k\phi^k$ is a polynomial of degree s.

Another way to construct Poincaré-invariant theory is to consider the interaction of Dirac field $\psi(x)$ and electromagnetic field $A_a(x)$. This interaction is described by the term $\int dx j^a(x)A_a(x)$ in action functional or by a similar term with three-dimensional integration in the Hamiltonian. The current $j(x)$ is divergence-free, therefore the action functional is invariant under gauge transformations $A_a \to A_a + \partial\epsilon$. Quantization of this action functional leads to quantum electrodynamics.

To quantize the theory of scalar field we approximate it by a theory with a finite number of degrees of freedom. This can be done in many ways. One of the possible ways is to replace the spatial variable with a discrete variable that runs over a cubic lattice (a grid) with lattice parameter a, derivatives with difference operators, and integration with summation over the lattice. This operation is one of the forms of momentum cutoff; to obtain a system with a finite number of degrees of freedom we should take a finite lattice (this is a form of infrared cutoff).

Following Sect. 2.8 we work in Hamiltonian formalism. Then the approximating Hamiltonian can be written in the form

$$\hat{H} = \sum_s \Big((C_1(a)\hat{p}_s^2 + C_2(a)(\nabla_a\hat{q}^s)^2 + \sum_k (g_k(a)(\hat{q}^s)^k\Big). \tag{3.15}$$

Here s runs over a cubic lattice with lattice parameter a, $\hat{p}_s$ and $\hat{q}^s$ obey CCR, ∇_a stands for the lattice analog of the gradient. We assume that the lattice is of finite size but the volume Ω covered by the lattice tends to infinity as $a \to 0$. It is convenient to take first the limit $\Omega \to \infty$ to get a translation-invariant theory and to take the limit $a \to 0$ afterward.

We can hope to obtain a Poincaré invariant theory in the limit for an appropriate choice of coefficients in (3.15). This choice can be based on the following idea. Fix a finite number of physical quantities (the particle mass and some scattering amplitudes) and find coefficients in (3.15) that give these quantities in the limit. One can conjecture that this choice leads to a well-defined Poincaré-invariant theory in the limit $a \to 0$. One should expect that this conjecture is true in the framework of perturbation theory if the degree s of the polynomial $P(\phi)$ in the action functional is ≤ 4. (One says in this case, that the theory is renormalizable.) In the case $s = 4$ one should fix the mass of the particle and two scattering amplitudes (physical coupling constants). We obtain a family of Poincaré-invariant local theories depending on three parameters.

To construct perturbation theory one can use the results of Sect. 2.8 taking quadratic Hamiltonian as $H(0)$.

For $s > 4$ this program does not go through (these theories are not renormalizable). However, one can apply these ideas to quantum electrodynamics (to the theory describing the interaction of quantized Dirac field and electromagnetic field). We obtain (in the framework of perturbation theory) a family of Poincaré-invariant theories depending on two parameters (physical mass and coupling constant).

Chapter 4
Deterministic Physical Theories from Geometric Viewpoint

We analyze deterministic physical theories taking as a starting point the set of states (geometric approach). For quantized classical theories we describe very convenient formalism where states are represented by functionals called L-functionals and give an application of this formalism to infrared problem in quantum electrodynamics. We follow mostly [11, 21, 26, 27, 29, 36].

Section 4.2. **contains reprinted material from the review** [10].

4.1 Introduction

In the present chapter, we analyze deterministic physical theories. We start with classical theories with a restricted set of observables (in other words, we assume that our devices can measure only a part of observables). We identify states that cannot be distinguished by "observable observables" (in other words we eliminate redundant states). We show that this procedure leads to a large class of theories that contains quantum theory (Sect. 4.2). These theories can be described in the geometric approach where the starting point is the set of states; we call them geometric theories (Sect. 4.3). In Sect. 4.4, we prove decoherence for a large class of geometric theories. (Our proof of decoherence is based on the consideration of a system in a random environment modeled by a random adiabatic Hamiltonian.) The standard formulas for probabilities can be obtained in quantum theory from decoherence; we generalize these considerations to theories defined in the geometric approach. In Sects. 4.5 and 4.6 we assume that a commutative group of time and spatial translations acts on the set of states. This allows us to define elementary excitations of translation-invariant stationary states. The elementary excitations of ground state should be interpreted as particles, and elementary excitations of arbitrary translation-invariant stationary state

A. Schwarz, *Quantum Mechanics and Quantum Field Theory in Algebraic and Geometric Approaches*, SpringerBriefs in Physics,
https://doi.org/10.1007/978-3-031-67915-5_4

should be interpreted as quasiparticles. To define the scattering of elementary excitations we need some additional conditions (like asymptotic commutativity or cluster property), but we do not need locality or Lorentz-invariance. We concentrate our attention on the inclusive scattering matrix, which is closely related to the inclusive cross-sections. This notion is necessary (and very natural) in the geometric approach, but it is useful also in the algebraic approach where inclusive scattering matrix can be expressed in terms of generalized Green functions introduced in Keldysh formalism of non-equilibrium statistical physics (see Chap. 2). Notice that the conventional scattering matrix cannot be defined for quasiparticles or when the theory does not have a particle interpretation; the inclusive scattering matrix still makes sense in these cases. In Sects. 4.7, 4.8, and 4.9 we represent states by functionals called L-functionals and show that the formalism of L-functionals is as convenient (and sometimes more convenient) as the formulation in terms of Hilbert space. As an example, we show that this formalism leads to a very simple solution to the infrared problem in quantum electrodynamics in the case when one can neglect the action of photons on electrons; this opens the way to a simple solution of the infrared problem in general case.

4.2 Classical Theory with a Restricted Set of Observables

Let us consider a classical theory with phase space M. In such a theory pure states are points of symplectic manifold M; mixed states are probability distributions on M; every mixed state can be uniquely represented as a mixture of pure states. Physical observables are real functions on M. An observable a specifies a vector field A on M as a Hamiltonian vector field with the Hamiltonian a. (Identifying a vector field with a first-order differential operator we can express A in terms of Poisson bracket: $Af = \{a, f\}$.) We assume that by integrating this vector field we obtain a one-parameter group $\sigma_A(t)$ of canonical transformations (symplectomorphisms) of M. This group acts also on mixed states and on observables describing the time evolution of these objects; the equation of motion for the density of probability distribution (Liouville equation) has the form $\frac{d\rho}{dt} = -\{a, \rho\}$ and the equation of motion for observables has the form $\frac{df}{dt} = \{a, f\}$.

Let us suppose that our devices are able to see only a part of observables. We assume that the set Λ of "observable observables" (of real functions on M that can be measured by our devices) is a linear space closed with respect to the Poisson bracket; we label this set by elements of Lie algebra denoted $\mathfrak{g}$. (The map $\gamma \to a_\gamma$ sending $\gamma \in \mathfrak{g}$ into $a_\gamma \in \Lambda$ is an isomorphism of Lie algebras $\mathfrak{g}$ and Λ.) Hamiltonian vector fields A_γ with Hamiltonians a_γ specify an action of Lie algebra $\mathfrak{g}$ on M. The assumption that vector fields A_γ generate one-dimensional subgroups means that this action comes from an action of simply connected Lie group G having $\mathfrak{g}$ as Lie algebra.

Notice that the considerations above can be applied to infinite-dimensional symplectic manifolds and to infinite-dimensional Lie algebras and Lie groups. However, in the infinite-dimensional case, our considerations here and in what follows are not quite rigorous (see Appendix A.1 for more rigorous discussion).

One defines the moment map μ of M to $\mathfrak{g}^\vee$ as a map $x \to \mu_x$ where $\mu_x(\gamma) = a_\gamma(x)$. (Here $x \in M, \gamma \in \mathfrak{g}$, and $\mathfrak{g}^\vee$ denotes the space of linear functionals on $\mathfrak{g}$.) This map is G-equivariant with respect to coadjoint action of G on $\mathfrak{g}^\vee$. For every state of the classical system (for every probability distribution ρ on M) we define a point $\nu(\rho) \in \mathfrak{g}^\vee$ as an integral of μ_x over $x \in M$ with respect to the measure ρ:

$$\nu(\rho) = \int_M \mu_x d\rho.$$

The point $\nu(\rho)$ belongs to the convex envelope $\mathcal{N}$ of $\mu(M) \subset \mathfrak{g}^\vee$. (The convex envelope of a subset E of topological vector space is defined as the smallest convex closed set containing E.)

The group G acts naturally on the space of classical states. It follows from G-equivariance of the moment map that the map ν is a G-equivariant map of this space into $\mathfrak{g}^\vee$ equipped with coadjoint action of G.

We say that two classical states (two probability distributions ρ and ρ') are equivalent if

$$\int_{x \in M} a_\gamma(x) d\rho = \int_{x \in M} a_\gamma(x) d\rho' \tag{4.1}$$

for every $\gamma \in \mathfrak{g}$. In other words, we say that two states are equivalent if calculations with these states give the same results for every Hamiltonian a_γ. (Our devices cannot distinguish these two states.)

We will derive the following statement:

Two states ρ and ρ' are equivalent iff $\nu(\rho) = \nu(\rho')$.

To give proof we notice that for every $\gamma \in \mathfrak{g}$

$$\nu(\rho)(\gamma) = \int_{x \in M} \mu_x(\gamma) d\rho = \int_{x \in M} a_\gamma(x) d\rho$$

and similarly

$$\nu(\rho')(\gamma) = \int_{x \in M} \mu_x(\gamma) d\rho' = \int_{x \in M} a_\gamma(x) d\rho'.$$

In the classical theory with Hamiltonians taken only from the set $\Lambda = \{a_\gamma\}$ where $\gamma \in \mathfrak{g}$ equivalent states should be identified (we eliminate redundant states; see the next section). The map ν induces a bijective map of the space of equivalence classes onto the set $\mathcal{N}$ obtained as a convex envelope of $\mu(M)$ ("quantum states"). The G-equivariance of the map ν means that the evolution of classical states agrees with the evolution of quantum states.

Let us apply our constructions to complex projective space $\mathbb{CP}$. We define this space as the sphere $||x|| = 1$ in complex Hilbert space $\mathcal{H}$ with identifications $x \sim \lambda x$ where $\lambda \in \mathbb{C}, |\lambda| = 1$. The group U of unitary operators acts transitively on $\mathbb{CP}$. There exists a unique (up to a constant factor) U-invariant symplectic structure on this space; this allows us to consider complex projective space as a homogeneous symplectic manifold.

One can construct a symplectic structure on $\mathbb{CP}$ considering it as a coadjoint orbit of the group U.

Let us suppose that we are able to observe only Hamiltonians of the form $a_C(x) = \langle x, Cx\rangle$ where C is a self-adjoint operator. The set Λ of such Hamiltonians is a Lie algebra with respect to the Poisson bracket; this Lie algebra is isomorphic to the Lie algebra $\mathfrak{g}$ of self-adjoint operators on $\mathcal{H}$ where the operation is defined as the commutator multiplied by i. The one-parameter group of unitary operators corresponding to the Hamiltonian a_C is given by the formula $\sigma_(t) = e^{-iCt}$. The moment map transforms a point x into a linear functional on the space of self-adjoint operators that maps an operator C into TrK_xC where K_x is a projection of $\mathcal{H}$ onto vector x, i.e. $K_x(z) = \langle x, z\rangle x$. (Recall that in our notations points of complex projective space are represented by normalized vectors.) The convex envelope of the image of the moment map consists of positive definite self-adjoint operators with unit trace (i.e. it consists of density matrices).

We see that by applying our general construction to complex projective space we obtain the conventional quantum mechanics. In this case, our considerations are close to Weinberg's "non-linear quantum mechanics" [38]. (Weinberg suggested considering the classical theory on $\mathbb{CP}$ as a deformation of quantum mechanics.)

We can say, that quantum mechanics can be interpreted as classical mechanics on complex projective space where some of the parameters describing classical state (=probability distribution on projective space) are hidden.

One can consider a more general case when the manifold M is a coadjoint orbit (an orbit of the group G in the space $\mathfrak{g}^\vee$). It is well known that such an orbit is a homogeneous symplectic manifold. Elements of the Lie algebra $\mathfrak{g}$ can be regarded as linear functions on $\mathfrak{g}^\vee$; let us denote by Λ the set of restrictions of these functions to the orbit. We consider classical theory on M assuming that only observables from the set Λ can be measured. Then eliminating redundant states we obtain a theory where the set of states is a convex envelope of the orbit. (In our case the moment map is simply the embedding of the orbit into $\mathfrak{g}^\vee$.)

We can generalize the considerations above assuming that the manifold M of pure states is not necessarily symplectic. Then one-parameter groups $\sigma_A(t)$ describing the time evolution should belong to some group $\mathcal{W}$ of transformations of M and the corresponding vector field A should belong to the Lie algebra $Lie(\mathcal{W})$ of the group $\mathcal{W}$. For every $A \in Lie(\mathcal{W})$ we should fix a function a on M. One can say that a pair (A, a) specifies an observable. We say that these data (the manifold M, the group $\mathcal{W}$ and pairs (A, a) where A belongs to the Lie algebra of the group $\mathcal{V}$ and $a = a(A)$ is a function on M) specify a "classical" theory. (For genuine classical theory M is a symplectic manifold, $\mathcal{W}$ is the group of symplectomorphisms, and A is a Hamiltonian vector field with the Hamiltonian a.) The group $\mathcal{W}$ acts in a natural way on vector fields A and on functions a; we assume that these actions are compatible.

If $\mathfrak{g}$ is a Lie subalgebra of $Lie(\mathcal{W})$ we can assume that a pair (A, a) specifies an "observable observable" if $A \in \mathfrak{g}$. To get rid of redundant states we say that two states are equivalent if calculations with these states give the same results for any a such that the corresponding A belongs to $\mathfrak{g}$ (see (4.1)). Factorizing with respect to this equivalence relation we obtain a set $\mathcal{N}$ that does not contain redundant states.

4.3 Geometric Theories

In Sect. 4.2 we considered classical theories with a restricted set of observables (the set of "observable observables") as well as more general "classical" theories. Eliminating redundant states (=identifying states that give the same values for all "observable observables") we obtain a convex set $\mathcal{N}$. The set $\mathcal{N}$ can be regarded as a new set of states. We will describe a class of theories where the set $\mathcal{N}$ is the primary object (geometric theories). We assume that one can consider a mixture of states; therefore we consider the set of states as a closed convex subset $\mathcal{N}$ of Banach space (or, more generally topological vector space) denoted by $\mathcal{L}$. (Instead of this set one can consider convex cone C of not necessarily normalized states where proportional states are identified.) Extreme points of $\mathcal{N}$ are called pure states; we assume that every element of $\mathcal{N}$ can be represented as a mixture of pure states. (It is well-known that every point of a compact convex subset of locally convex topological vector space can be represented as a mixture of its extreme points. From the other side, a bounded closed convex subset of reflexive Banach space is weakly compact.)

We fix a subgroup $\mathcal{V}$ of the group of automorphisms of $\mathcal{N}$.

An observable in a geometric theory is specified by an element A of the Lie algebra of the group $\mathcal{V}$ and a linear functional a on $\mathcal{L}$ corresponding to A. We say that $a = a(A)$ is a Hamiltonian function and A is a "Hamiltonian" (in quotation marks). The set of observables is denoted by Λ. The group $\mathcal{V}$ acts naturally on $A \in Lie(\mathcal{V})$ (according to adjoint representation) and on the linear functional a corresponding to A; these two actions should be compatible (if $g \in \mathcal{V}$ transforms A into A' then it should transform $a = a(A)$ into $a' = a(A')$).

For every $A \in Lie(\mathcal{V})$ we define the evolution operator $\sigma_A(t)$ as a solution of the equation of motion:

$$\frac{d\sigma_A(t)}{dt} = A\sigma_A(t) \tag{4.2}$$

We assume that there exists a solution $\sigma_A(t) \in \mathcal{V}$. (In other words, every element of the Lie algebra $Lie(\mathcal{V})$ specifies an evolution operator). The element A is $\sigma_A(t)$-invariant, hence Hamiltonian function a is also invariant with respect to $\sigma_A(t)$ (the latter statement is equivalent to the condition $a(Ax) = 0$.)

One can consider also a more general case when A in (4.2) depends on t.

In textbook quantum mechanics $\mathcal{N}$ consists of density matrices (positive definite self-adjoint operators with unit trace acting in Hilbert space $\mathcal{H}$). The equation of motion in the case of time-independent Hamiltonian $\hat{A}$ has the form

$$\frac{dK}{dt} = AK = -i(\hat{A}K - K\hat{A}). \tag{4.3}$$

The space $\mathcal{L}$ consists of all operators having trace (belonging to the trace class). The cone C consists of positive definite trace class operators. The group $\mathcal{V}$ is isomorphic to the group of unitary operators in $\mathcal{H}$; these operators act on $\mathcal{L}$ by the formula $K \to U^{-1}KU$.

Observables (A, a) correspond to self-adjoint operators $\hat{A}$. The operator A (the "Hamiltonian") is defined as a commutator with $\hat{A}$ multiplied by i. (In this case the "Hamiltonian" is called Liouvillean.) The Hamiltonian function $a(K)$ is defined as a linear functional on $\mathcal{L}$ given by the formula $a(K) = Tr\hat{A}K$.

In the algebraic approach to quantum theory, the starting point is a $*$-algebra $\mathcal{A}$ (a unital associative algebra with involution *).The cone C consists of positive linear functionals on $\mathcal{A}$. (One says that a linear functional ω is positive if $\omega(x^*x) \geq 0$ for every $x \in \mathcal{A}$). The set $\mathcal{N}$ consists of positive functionals obeying the normalization condition $\omega(1) = 1$. The space $\mathcal{L} = \mathcal{A}^\vee$ consists of all linear functionals on $\mathcal{A}$. The group $\mathcal{V}$ can be interpreted as the group of involution-preserving automorphisms of $\mathcal{A}$ acting naturally on the dual space $\mathcal{L}$.

An observable (A, a) corresponds to a self-adjoint element $\hat{A}$ of $\mathcal{A}$. The Hamiltoian function $a(K)$ where $K \in \mathcal{L}$ is defined by the formula $a(K) = K(\hat{A})$. The evolution operator $\sigma_A(t)$ acts on $\mathcal{A}$ as conjugation with $e^{i\hat{A}t}$; this action induces an action of $\sigma_A(t)$ on $\mathcal{L} = \mathcal{A}^\vee$. The "Hamiltonian" A specifying the evolution of elements of acts on $\mathcal{A}$ as a commutator with $\hat{A}$ (up to a factor of i); this action induces an action of A on $\mathcal{L}$.

If $\mathcal{A}$ is a topological algebra one should assume that all functionals and operators are continuous. If $\mathcal{A}$ is a C^*-algebra then $e^{i\hat{A}t}$ is a well defined unitary element of $\mathcal{A}$, hence the equation of motion has a solution and the evolution operator $\sigma_A(t)$ is well defined.

In the situation of Sect. 4.2 classical theory with a restricted set of observables is equivalent to geometric theory. The set of states $\mathcal{N}$ should be identified with convex envelope $\mathcal{N}$ of $\mu(M)$, the group G plays the role of $\mathcal{V}$. The pairs (A_γ, a_γ) where $\gamma \in \mathfrak{g}$ play the role of observables.

The considerations of Sect. 4.2 can be generalized as follows. We say that there exist redundant states in a geometric theory if one can find such states $x, y \in \mathcal{N}$ that for every observable (A, a) we have $a(x) = a(y)$ (there are no observables that allow us to distinguish these states). In this case, it is useful to work with theory without redundant states (to eliminate redundant states). To construct such a theory we introduce an equivalence relation in $\mathcal{L}$ saying that $x \sim y$ if $a(x) = a(y)$ for every observable (A, a). In the new theory the set of states $\mathcal{N}'$ is defined as a set of equivalence classes in $\mathcal{N}$. The group $\mathcal{V}$ acts on $\mathcal{L}'$ (on the space of equivalence classes in $\mathcal{L}$); its elements can be regarded as automorphisms of $\mathcal{N}'$. The observables descend to $\mathcal{L}'$.

Section 4.2 gives a concrete realization of this procedure in the case when we are starting with classical theory. Let us check that one can obtain every geometric theory eliminating redundant states in a "classical" theory. (The definition of "classical" theory is given at the end of Sect. 4.2. One can say that a "classical" theory is a geometric theory where every state has a unique representation as a mixture of pure states.) Starting with a geometric theory we construct a "classical" theory taking as M the set of extreme points of $\mathcal{N}$. The automorphisms of $\mathcal{N}$ act on M, hence we can consider $\mathcal{V}$ as a subgroup of the group of transformations of M and elements of its Lie algebra as vector fields on M. Linear functionals $a(A)$ specify functions on $M \subset \mathcal{L}$. We got a "classical" theory; using the assumption that every element of $\mathcal{N}$ is a mixture of pure states we conclude that the geometric theory can be obtained from this "classical" theory.

4.4 Decoherence. Probabilities

Let us show that under some conditions one can prove a generalization of decoherence for a geometric theory. The decoherence in quantum mechanics can be used to derive the standard formulas for probabilities; similar considerations can be applied to a geometric theory.

Decoherence appears if we place a system into a random environment. We suppose that the evolution in geometric theory is described by a "Hamiltonian" A; the random environment is modeled by an adiabatic "Hamiltonian" $A(g(t))$ where $A(g)$ is a family of random "Hamiltonians" such that $A(0) = A$. Let us assume that all zero modes of the "Hamiltonian" A are robust; this means that for every $x \in \mathcal{L}$ obeying $Ax = 0$ (for every zero mode of A) and sufficiently small g, there exists a zero mode $x(g)$ of $A(g)$ (a vector $x(g)$ obeying $A(g)x(g) = 0$) that depends continuously on g. Decoherence means that the interaction with a random environment makes the evolution of states except robust zero modes unpredictable. In standard quantum mechanics, where the "Hamiltonian" A is a commutator of the density matrix K with the Hamiltonian $\hat{A}$ (up to a numerical factor) the matrix K is a zero mode of A if it commutes with $\hat{A}$. All zero modes of A are robust if all eigenvalues of $\hat{A}$ are simple. (In this case, every zero mode of A can be represented as $f(\hat{A})$ where f is some function and the zero mode of $A(g)$ that depends continuously on g can be written as $f(\hat{A}(g))$.) In $\hat{A}$-representation only diagonal elements of the density matrix K are predictable.

We assume that the set of states $\mathcal{N}$ is bounded. It follows that for every "Hamiltonian" A the norms of evolution operators $\sigma_A(t)$ are bounded from above by a constant that does not depend on t. This means that all eigenvalues of A are purely imaginary. If the space $\mathcal{L}$ were finite-dimensional we could say that A is diagonalizable; in an infinite-dimensional case, this is not necessarily true, but still, we can say that A does not have non-trivial Jordan cells.

We assume that the "Hamiltonian" A is diagonalizable, all zero modes are robust, and all non-zero eigenvalues $\epsilon_j \in i\mathbb{R}$ are simple; corresponding eigenvectors are denoted ψ_j. These eigenvectors can be deformed into eigenvectors of $A(g)$ for small g. More precisely, for $|g| < \delta_j$ we can construct vectors $(\psi_j(g))$ that depend continuously on g in such a way that

$$A(g)\psi_j(g) = \epsilon_j(g)\psi_j(g) \tag{4.4}$$

where $\psi_j(0) = \psi_j$. The assumption that zero modes are robust implies that for zero modes ψ_j we can assume that $\epsilon_j(g) \equiv 0$.

We model the interaction with the environment by random perturbation $A(g(t))$. Then in the adiabatic approximation

$$\sigma_A(t)\psi_j = e^{\rho_j(t)}\psi_j(g(t)), \tag{4.5}$$

where $\frac{d\rho_j}{dt} = \epsilon_j(g(t))$. (In adiabatic approximation we can neglect the derivative $\dot{g}(t)$. It follows that the right-hand side of (4.5) obeys the equations of motion in this approximation. We assume that $g(t)$ takes values in a small neighborhood of $g = 0$.)

The evolution of (robust) zero modes of A is predictable (at least in the case when the set $\mathcal{R}$ of states that are zero modes has a finite number of extreme points, or, more generally, when the set of extreme points is discrete). To check this we notice that the zero modes of A evolve to zero modes of $A(g(t))$; in other words, the evolution operator maps $\mathcal{R}$ onto the set $\mathcal{R}(t)$ of zero modes of $A(g(t))$. Noticing that extreme points of $\mathcal{R}$ are mapped to extreme points of $\mathcal{R}(t)$ we obtain that the evolution of zero modes is predictable. (Recall that we assume that all zero modes are robust.). The evolution of other eigenvectors of A is unpredictable. Imposing some conditions on the random "Hamiltonian" $A(g(t))$ one can prove that on average the random phase factors $e^{\rho_j(t)}$ vanish unless ϕ_j is a zero mode. Let us sketch the proof of this fact if $g(t) = \alpha t, \alpha \to 0$, and the probability distribution on "Hamiltonians" comes from a probability distribution on the parameter g. In this case, the expectation value of the random phase factor can be written in the form

$$\int d\mu e^{\frac{1}{\alpha}\rho_j(g)}$$

where $\frac{d\rho_j}{dg} = \epsilon_j(g)$.

Assuming that the measure μ (the probability distribution on g) is absolutely continuous and taking into account that $\rho_j(g)$ is purely imaginary we obtain that the expectation value tends to zero as $\alpha \to 0$.

If the evolution of vector $\Psi = \sum c_i\psi_i$ is governed by random adiabatic "Hamiltonian" $A(g(t)$ with $g(0) = g(T) = 0$ we can say that at the moment T the coefficients c_i at the moment T are the same as in the moment 0 for zero modes ψ_i; for all other eigenvectors the expectation values of the coefficients c_i vanish. This statement generalizes the decoherence of quantum mechanics.

Let us denote by $KerA$ the space of zero modes of A and by P the projection $P : \mathcal{L} \to KerA$ sending all eigenvectors of A that are not zero modes to zero. (Recall that we assume that all zero modes of A are robust.)

Notice that the projection P can be represented by the formula

$$Px = \lim_{T\to\infty} \frac{1}{T} \int_0^T dt\sigma_A(t)x$$

It follows from this formula that $P\mathcal{N} \subset \mathcal{N}$.

To calculate probabilities of the observable (A, a) in the state x we should represent the zero mode Px as a mixture of pure zero modes of extreme points of the set $\mathcal{R}$ consisting of zero modes belonging to $\mathcal{N}$:

$$Px = \sum p_k z_k. \tag{4.6}$$

We interpret p_k as the probability of finding the value $a(z_k)$ measuring the observable (A, a). (We assume that the numbers $a(z_k)$ are different. If this condition is not satisfied we should calculate the probability to obtain the value α summing all p_k with $a(z_k) = \alpha$.) This interpretation generalizes the standard prescription of textbook quantum mechanics.

Recall that in quantum mechanics, we take A as a commutator of a self-adjoint operator $\hat{A}$ with density matrix K multiplied by i and define $a(K) = Tr\hat{A}K$. Notice, that the representation (4.6) of zero mode as a mixture of pure zero modes, in general, is not unique. However, in conventional quantum mechanics, this representation is unique in the case when A corresponds to an operator $\hat{A}$ having simple eigenvalues. In $\hat{A}$-representation a basis of eigenvectors of A consists of matrices having only one non-zero entry equal to 1. Zero modes of A are diagonal matrices, and all of them are robust. Every diagonal matrix can be represented uniquely as a mixture of pure zero modes of A and the coefficients of this representation (diagonal entries of the matrix K in $\hat{A}$-representation) are quantum probabilities.

Let us illustrate the above considerations in the situation of Sect. 4.2 assuming that the symplectic manifold M is a coadjoint orbit of Lie group G and the set of "observable observables" is identified with the Lie algebra $\mathfrak{g}$ of G. (To every element $\gamma \in \mathfrak{g}$ we assign a pair (A_γ, a_γ) where a_γ is the restriction to the orbit of linear functional on $\mathfrak{g}^\vee$ specified by γ and the "Hamiltonian" A_γ comes from coadjoint representation of $\mathfrak{g}$.) Then the space of states of the theory obtained by the elimination of redundant states is a convex envelope $\mathcal{N}$ of the orbit; pure states belong to the orbit.

For a compact Lie group, the coadjoint representation can be identified with adjoint representation. Without loss of generality, we can assume that the "Hamiltonian" A_γ corresponds to an element γ belonging to Cartan subalgebra $\mathfrak{h}$. (This follows from the fact that all Cartan subalgebras are conjugate.) Using commutativity of Cartan subalgebra we obtain that elements of $\mathfrak{h}$ are zero modes of A_γ. Elements of $\mathfrak{h}$ belonging to an orbit are pure zero modes. We assume that γ is a regular element (i.e.the set of elements commuting with it is a commutative subalgebra); then all zero modes of A_γ belong to $\mathfrak{h}$; all zero modes are robust. (To prove this we notice that the set of regular elements is open and that the Cartan subalgebra corresponding to a regular element depends continuously on this element.)) The operator P can be interpreted as an orthogonal projection of $\mathfrak{g}$ onto Cartan subalgebra $\mathfrak{h}$. To calculate probabilities in the state x we represent the zero mode Px as a mixture of pure robust zero modes of A_γ. Notice that pure robust zero modes of A_γ can be identified with the intersection points of the orbit with the Cartan subalgebra $\mathfrak{h}$. To calculate the number of points in the intersection of an orbit with $\mathfrak{h}$ one can use the theorem that this intersection can be identified with an orbit of the Weyl group in $\mathfrak{h}$.

If G is a unitary group then the elements of its Lie algebra considered as matrices are regular iff all eigenvalues are distinct. Let us consider zero modes (= stationary states) of the Hamiltonian" A_γ corresponding to regular γ; they can be identified with elements of the Cartan sub-algebra containing γ. Probabilities in the state x can be calculated as coefficients in the representation (4.6) of the zero mode Px as a mixture of pure zero modes. As we know in the case when the orbit can be identified with complex projective space the construction of Sect. 4.2 gives conventional quantum mechanics. In this case, the representation in the form (4.6) is unique (as should be in quantum theory in the case of the simple spectrum). For all other orbits the representation in the form (4.6) is not unique. This means that corresponding physical theories cannot be described in the framework of conventional quantum mechanics.

4.5 Particles as Elementary Excitations

To develop scattering theory in the geometric approach we need time translations T_τ and spatial translations $T_{\mathbf{a}}$ acting in the cone of states $C \subset \mathcal{L}$. These notions allow us to define elementary excitations of translation-invariant stationary state ω.

We define elementary space $\mathfrak{h}$ as a space $\mathcal{S}$ of complex vector-valued smooth fast decreasing functions on $\mathbb{R}^d$ where spatial translations act as shifts of argument: $(T_{\mathbf{a}} f)(\mathbf{x}) = f(\mathbf{x}+\mathbf{a})$ and time translations T_τ are unitary operators commuting with spatial translations. (A scalar product of two functions taking values in $\mathbb{C}^r$ is specified by the standard formula $\langle f, g\rangle = \sum_1^r \int d\mathbf{x} \bar{f}_k(\mathbf{x}) g_k(\mathbf{x})$.)

In momentum representation spatial translation T_a is defined as multiplication by $e^{i\mathbf{ap}}$ and time translation T_τ can be represented as multiplication by $e^{-iE(\mathbf{p})\tau}$ where $E(\mathbf{p})$ is a Hermitian $r \times r$ matrix.

We say that an elementary space is admissible if the matrix $E(\mathbf{p})$ is positive definite.

An elementary excitation of the state ω is specified by a map σ of elementary space $\mathfrak{h}$ into the cone C commuting with spatial and temporal translations and bounded linear operators $L(\phi), \phi \in \mathfrak{h}$ acting in the space $\mathcal{L}$ and obeying

$$L(\phi)\omega = \sigma(\phi) \tag{4.7}$$

Notice that we do not suppose that linear operators L depend linearly on ϕ; it is natural to assume that their dependence on ϕ is quadratic, but this assumption will not be necessary to define the scattering matrix.

The operators $L(\phi)$ should satisfy some additional conditions; not very precisely one can say that $L(\phi)$ and $L(\psi)$ most commute when the supports of ϕ and ψ in coordinate representation are far away. This assumption can be made precise in various ways.

The space of excitations of the state ω can be defined as the smallest closed linear subspace of $\mathcal{L}$ containing ω and invariant with respect to operators $L(\phi)$.

If the elementary space is admissible we say that the elementary excitation is a particle, if it is not admissible we say that this excitation is a quasiparticle.

In relativistic theory the action of the group of translations on the cone C should be extended to the action of the Poincaré group and the state ω should be Poincaré-invariant. Similarly, the Poincaré group should be represented by unitary transformations of an elementary space $\mathfrak{h}$. A relativistic particle is defined as a map of admissible elementary space equipped with an irreducible representation of Poinaré group into the cone C; this map should commute with actions of Poincaré group in $\mathfrak{h}$ and C.

4.6 Scattering. Inclusive Scattering Matrix

We would like to consider the scattering of elementary excitations. It seems that one cannot define the conventional scattering matrix in the geometric approach, however, there exists a very natural definition of the inclusive scattering matrix. This definition is based on the consideration of operators $L(f, \tau)$ specified by the formula

$$L(f, \tau) = T_\tau(L(T_{-\tau} f)) = T_\tau(L(T_{-\tau} f)T_{-\tau}$$

where $L(f)$ denotes the operator entering the definition of elementary excitation (see (4.7).

Using the assumption that ω is a stationary state and the assumption that σ commutes with temporal translations it is easy to check that $L(f, \tau)\omega$ does not depend on τ, hence

$$\dot{L}(f, \tau)\omega = 0 \tag{4.8}$$

To define (inclusive) scattering matrix we fix translation-invariant stationary element α of the dual space $\mathcal{L}^\vee$. Then the (inclusive) scattering matrix (more precisely, (α, ω) scattering matrix) is defined as a functional

$$S_{n',n}(g_1', ..., g_{n'}', g_1, ..., g_n) = \lim_{\tau' \to +\infty, \tau \to -\infty} \langle \alpha | L(g_1', \tau')...L(g_{n'}', \tau')L(g_1, \tau)...L(g_n, \tau)|\omega\rangle \tag{4.9}$$

We assume that the functions $(g_1, ..., g_n)$ *do not overlap as well as the functions* $(g_1', ..., g_{n'}')$.

It follows that essential supports of functions $T_\tau g_j$ are far away for $\tau \to -\infty$ (see Appendix A.3). This implies that operators $L(g_i, \tau)$ and $L(g_j, \tau)$ commute in this limit. We conclude that $S_{n,n'}$ is symmetric with respect to $g_1, ..., g_n$; similarly, we can prove that it is symmetric with respect to $g'_1, .., g'_{n'}$.

To prove the existence of the limit as $\tau \to -\infty$ in (4.9) we assume that

$$||[\dot{L}(g_i, \tau), L(g_j, \tau)]|| < c(\tau) \tag{4.10}$$

where $c(\tau)$ is a summable function. Then the derivative of the function under the lim sign with respect to τ is also summable; the existence of the limit follows from this fact. To estimate this derivative we use the Leibniz rule. We obtain n summands, and to prove that every summand is summable we apply (4.10) to transfer the factor with time derivative to the rightmost place and use (4.8). (Similar arguments were used many times in Chap. 2.) One should expect that the condition (4.10) is satisfied if functions g_i, g_j do not overlap. *In what follows we assume that this statement is correct.* (This is one of rigorous versions of conditions we imposed on operators $L(\phi)$.)

To justify the definition (4.9) we introduce the notion of scattering state (or, more precisely, of in-state) by the formula

$$\Lambda(f_1, \cdots, f_n| - \infty) = \lim_{\tau_1 \to -\infty, \cdots, \tau_n \to -\infty} \Lambda(f_1, \tau_1, \cdots, f_n, \tau_n) \tag{4.11}$$

where

$$\Lambda(f_1, \tau_1, ..., f_n, \tau_n) = L(f_1, \tau_1)...L(f_n, \tau_n)\omega.$$

We say that (4.11) is an in-state. It follows from the considerations above that the limit in (4.11) exists if functions $f_1, ..., f_n$ do not overlap and $\tau_1 = ...\tau_n$. A slight modification of these considerations allows us to prove that this limit exists also in the case when the arguments τ_i tend to $-\infty$ independently.

For large negative τ the state

$$T_\tau \Lambda(f_1, \cdots, f_n| - \infty)$$

can be described as a collection of particles with wave functions $T_\tau f_i$. To prove this fact we use the formulas

$$T_\tau(L(f, \tau')) = T_{\tau+\tau'} L(T_{-\tau'} f) T_{-\tau-\tau'} = L(T_\tau f, \tau + \tau'),$$

$$T_\tau \Lambda(f_1, \cdots, f_n| - \infty) = \Lambda(T_\tau f_1, \cdots, T_\tau f_n| - \infty).$$

If functions $f_1, ..., f_n$ do not overlap we have a collection of distant particles for $\tau \to -\infty$. This fact allows us to say that the state $T_\tau \Lambda(f_1, \cdots, f_n| - \infty)$ describes a collision of particles with wave functions $(f_1, \cdots, f_n)$. A number describing this state in the limit $\tau \to +\infty$ can be interpreted as a matrix element of the inclusive scattering matrix. This remark justifies the definition (4.9) (notice that the operators $L(g'_i, \tau')$ in this definition should be considered as detectors).

Let us introduce notations

$$L_{out}(g) = \lim_{\tau' \to +\infty} L(g, \tau'), L_{in}(g) = \lim_{\tau \to -\infty} L(g, \tau). \qquad (4.12)$$

In these notations the definition (4.9) can be written as:

$$S_{n',n}(g'_1, ..., g'_{n'}, g_1, ..., g_n) = \langle \alpha | L_{out}(g'_1)...L_{out}(g'_{n'}) | \Lambda(g_1, ..., g_n | -\infty) \rangle \quad (4.13)$$

or as

$$S_{n',n}(g'_1, ..., g'_{n'}, g_1, ..., g_n) = \langle \alpha | L_{out}(g'_1)...L_{out}(g'_{n'}) L_{in}(g_1)...L_{in}(g_n) | \omega \rangle \quad (4.14)$$

Let us formulate some conditions for existence of limits in (4.12). One can check that the limit $L(g, \tau)\Lambda$ as $\tau \to -\infty$ exists if $\Lambda = \Lambda(g_1, ..., g_n | -\infty)$ and functions $g, g_1, ..., g_n$ do not overlap; in this case $L_{in}\Lambda = \Lambda(g, g_1, ..., g_n | -\infty)$.

To prove this statement we notice that

$$\Lambda(g, g_1, ..., g_n | -\infty) = \lim_{\tau, \tau_i \to \infty} \Lambda(g, \tau, g_1, \tau_1, ..., g_n, \tau_n)$$

As we noticed one can assume that the arguments $\tau, \tau_1, ..., \tau_n$ tend to $-\infty$ independently, hence we can take the limit $\tau_i \to -\infty$ first and after that take the limit $\tau \to -\infty$. This gives us the statement we need as well as similar statement for operators L_{out}.

One should expect that operators L_{in}, L_{out} are well defined on a dense open subset of the space of excitations of the state ω.

In the algebraic approach, the starting point is a $*$-algebra $\mathcal{A}$; time shifts and spatial shifts come from automorphisms of $\mathcal{A}$. We can construct the data of the geometric approach identifying the cone C of states with the cone of positive functionals on $\mathcal{A}$ (of linear functionals obeying $\omega(A^*A) \geq 0$) and the space $\mathcal{L}$ with the space of continuous linear functionals on $\mathcal{A}$. Fixing a translation-invariant stationary element $\alpha \in \mathcal{A} \subset \mathcal{L}^\vee$ and translation-invariant stationary state $\omega \in C$ we can construct (inclusive) scattering matrix using (4.9). More precisely, we consider an elementary excitation of ω as a linear map Φ of the elementary space $\mathfrak{h}$ into pre-Hilbert space $\mathcal{H}$ obtained by GNS-construction applied to the state ω. We assume that $\Phi(\phi) = \hat{B}(\phi)\theta$ where $\theta \in \mathcal{H}$ stands for the cyclic vector corresponding to the state ω and $B(\phi) \in \mathcal{A}$ depends linearly on ϕ. Then the state $\sigma(\phi)$ corresponding to the vector $\Phi(\phi)$ can be written in the form (4.7) with $L(\phi) = \tilde{B}(\phi)B(\phi)$ where we use notations $(\tilde{B}\sigma)(x) = \sigma(B^*(x), (B\sigma)(x) = \sigma(xB)$. Imposing the condition of asymptotic commutativity on the algebra $\mathcal{A}$ and choosing appropriate operators $B(\phi)$ one can prove that operators $L(\phi)$ obey the conditions we need to construct the inclusive scattering matrix.

Operators $B(f, \tau) = T_\tau B(T_\tau f) T_{-\tau}$ can be used to construct Møller matrices, conventional scattering matrix, *in*- and *out*-operators (see Sect. 2.3). The relations $L(f, t) = \tilde{B}(f, t) B(f, t)$ and $\lim_{\tau \to +\infty} B(f, t) = a^+_{out}(f)$ can be used to relate inclusive scattering matrix to inclusive cross-sections in the case when α is the unit element of the algebra $\mathcal{A}$ (see the end of Sect. 2.3 for details).

For the expression of of inclusive scattering matrix in algebraic approach in terms of generalized Green's functions see Sect. 2.7.

4.7 L-functionals

The main idea of the geometric approach is to take as a starting point the set of states. It is important to emphasize that working with the set of states is as convenient as working with Hilbert spaces (and sometimes more convenient). In this section, we illustrate this by representing states by functionals called L-functionals. (This representation was suggested in [21]. The formalism of L-functionals was rediscovered in less transparent form under the name of thermo-field dynamics (TFD); see [7].)

Let us quantize a classical theory with an infinite number of degrees of freedom. The Hamiltonian of the classical theory is a functional on the phase space with coordinates p_k, q^k having standard Poisson brackets; after quantization, we obtain operators $\hat{p}_k, \hat{q}^k$ obeying canonical commutation relations (CCR). It will be convenient to work with CCR for operators $\hat{a}(k), \hat{a}^*(k)$ that can be regarded as generalized functions of continuous and discrete parameters. In other words, we are working with operators $\hat{a}(f) = \int dk f(k) \hat{a}(k), \hat{a}^*(f) = \int dk f(k) \hat{a}^*(k)$ where f runs over the space of test functions $\mathcal{E}$ considered as pre-Hilbert space. The integral over k is considered as an integral over continuous parameters and a sum over discrete parameters. For definiteness, we assume that $\mathcal{E}$ is the space $\mathcal{S}$ of smooth fast-decreasing functions taking values in $\mathbb{C}^r$.

The CCR can be written in the form

$$[\hat{a}(f), \hat{a}(g)] = [\hat{a}^*(f), \hat{a}^*(g)] = 0, [\hat{a}(\bar{f}), \hat{a}^*(g)] = \hbar \langle f, g \rangle \tag{4.15}$$

In the case of an infinite number of degrees of freedom, there exist representations of CCR that are not equivalent to the standard Fock representation where $\hat{a}(f), \hat{a}^*(f)$ can be interpreted as annihilation and creation operators (see Sect. 1.8). In Hilbert space approach vectors and density matrices in all representation spaces can be regarded as states of the theory at hand. In the geometric approach, we can represent the states as functionals

$$\mathbf{L}_K(f) = Tr \hat{W}_f K. \tag{4.16}$$

where $\hat{W}_f = e^{-\hat{a}^*(f)}e^{\hat{a}(\bar{f})}$. It is easy to verify that the functional (4.16) is well-defined for a density matrix K in any representation of CCR. (The operator $-i\hat{a}^*(f) + i\hat{a}(\bar{f})$ is self-adjoint; this is a rigorous form of the statement the operator $\hat{a}^*(f)$ is Hermitian conjugate to the operator $\hat{a}(\bar{f})$. Using the fact that f is square integrable we obtain that up to a finite factor the operator $\hat{W}_f$ coincides with unitary operator $e^{-\hat{a}^*(f)+\hat{a}(\bar{f})}$. It follows that the operator $\hat{W}_f K$ belongs to the trace class.)

One can say that when working with functionals $\mathbf{L}$ we consider all representations of CCR simultaneously.

To emphasize that $\mathbf{L}_K$ does not depend analytically on f we use the notation $\mathbf{L}_K(\bar{f}, f)$ or $\mathbf{L}_K(f^*, f)$.

The condition $TrK = 1$ leads to normalization condition $\mathbf{L}_K(0, 0) = 1$.The density matrix K is a self-adjoint operator hence $\mathbf{L}_K(\bar{f}, f) = (\mathbf{L}_K(-\bar{f}, -f))^*$. The operator K is positive definite, this implies some positivity conditions on $\mathbf{L}_K(\bar{f}, f)$. We denote by $\mathcal{L}$ the vector space of non-linear functionals $\mathbf{L}(\bar{f}, f)$ and by $\mathcal{N}$ the subset of $\mathcal{L}$ satisfying the conditions above. (To be more precise we should work with the exponential form $\mathcal{W}$ of the Weyl algebra that can be defined as a closed subalgebra of the Banach algebra of bounded linear operators in Fock space generated by operators W_f. We can define $\mathcal{L}$ as the space of continuous linear functionals on $\mathcal{W}$; linear combinations of operators W_f are dense in $\mathcal{W}$, hence a linear functional $L \in \mathcal{L}$ is specified by numbers $L(W_f)$. The set $\mathcal{N}$ can be identified with the set of positive linear functionals on $\mathcal{W}$ represented by non-linear functionals $L(W_f)$.)

We will use the geometric approach assuming that $\mathcal{N}$ is the set of states. Discarding the normalization condition $\mathbf{L}(0, 0) = 1$ we obtain the cone C of non-normalized states; we will work with this cone.

It is easy to check that

$$\tilde{b}^*(f)\mathbf{L}_K = \mathbf{L}_{K\hat{a}(f)}, \quad \tilde{b}(f)\mathbf{L}_K = \mathbf{L}_{K\hat{a}^*(f)}, b(f)\mathbf{L}_K = \mathbf{L}_{\hat{a}(f)K}, \quad b^*(f)\mathbf{L}_K = \mathbf{L}_{\hat{a}^*(f)K} \tag{4.17}$$

where

$$b(f) = -\hbar c_2^*(f) + c_1(f), \quad b^*(f) = -c_2(f), \tag{4.18}$$

$$\tilde{b}^*(f) = \hbar c_1^*(f) - c_2(f), \quad \tilde{b}(f) = c_1(f), \tag{4.19}$$

$c_1^*(\bar{f})$ is a multiplication operator by $\bar{f}$, $c_2^*(f)$ is a multiplication operator by f, and $c_1(\bar{f})$, $c_2(f)$ are variational derivatives with respect to $\bar{f}$ and f.

More generally, to an operator $\hat{A}$ in a representation space of CCR we assign two operators A and $\tilde{A}$ in $\mathcal{L}$ obeying $AL_K = L_{\hat{A}K}$ and $\tilde{A}L_K = L_{K\hat{A}^*}$. Notice that the map $A \to \tilde{A}$ is antilinear, $\widetilde{AB} = \tilde{A}\tilde{B}$, $A\tilde{B} = \tilde{B}A$. It is easy to verify that the sets $\mathcal{N}$ and C are invariant under the operators of the form $e^{tA}e^{t\tilde{A}}$. In other words, if the equation of motion has the form

$$\frac{d\sigma}{dt} = (A + \tilde{A})\sigma,$$

then $\mathcal{N}$ and C are invariant under the evolution operator specified by a "Hamiltonian" $A+\tilde{A}$.

Unfortunately, a simple description of the set $\mathcal{N}$ does not exist. For our goals it is convenient to consider this set as a minimal closed subset of $\mathcal{L}$ invariant under all operators of the form $\exp(A+\tilde{A})$ and containing Gaussian functionals $e^{-\langle f,Sf\rangle}$ where S is a positive definite operator.

Quantizing classical Hamiltonian we obtain a formal expression

$$\hat{H}=\sum_{m,n}\int \Gamma_{m,n}(k_1,...k_m|l_1,...,l_n)\hat{a}^*(k_1)...\hat{a}^*(k_m)\hat{a}(l_1)...\hat{a}(l_n)d^mkd^nl. \tag{4.20}$$

We presented it in the normal form (i.e. all creation operators are moved to the left). In many interesting cases, the formal expression (4.20) does not define a self-adjoint operator in Fock space, but the corresponding equation of motion in the space $\mathcal{L}$ of functionals $\mathbf{L}(\bar{f},f)$ is well-defined. To write down this equation we calculate the operator corresponding in $\mathcal{L}$ to the commutator of $\hat{H}$ with K using (4.17). We obtain

$\frac{d\mathbf{L}}{dt}=(H+\tilde{H})\mathbf{L}$ where

$H=\frac{1}{i\hbar}(\sum_{m,n}\int\Gamma_{m,n}(k_1,...k_m|l_1,...l_n)b^*(k_1)...b^*(k_m)b(l_1)...b(l_n)d^mkd^nl$

$\tilde{H}=\frac{-1}{i\hbar}(\sum_{m,n}\int\bar{\Gamma}_{m,n}(k_1,...k_m|l_1,...l_n)\tilde{b}^*(k_1)...\tilde{b}^*(k_m)\tilde{b}(l_1)...\tilde{b}(l_n))d^mkd^nl$

Notice that it follows from (2.16) and (4.19) that the equation of motion for $\mathbf{L}$ has finite limit as $\hbar\to 0$.

Let us consider in more detail quadratic Hamiltonian

$$\hat{H}(0)=\int\epsilon(k)\hat{a}^*(k)\hat{a}(k)dk. \tag{4.21}$$

Here $k=(\mathbf{k},s)$ where $\mathbf{k}$ runs over $\mathbb{R}^d$, and s runs over a finite set X, integration over k is understood as integration over $\mathbf{k}$ and summation over discrete intex s. This Hamiltonian commutes with momentum operator $\mathbf{P}=\int\mathbf{k}\hat{a}^*(k)\hat{a}(k)dk$, hence it is translation-invariant.

It is easy to verify that the functionals of the form

$$\mathbf{L}_n=e^{-\int\bar{f}(k)n(kf(k)dk}, \tag{4.22}$$

are translation-invariant stationary states of the corresponding "Hamiltonian"

$$H(0)=i\int\epsilon(k)(c_1^*(k)c_1(k)-c_2^*(k)c_2(k))dk. \tag{4.23}$$

Notice that the states (4.22) can be characterized as normalized solutions of equations

$$A(k)\mathbf{L}_n = 0, \tilde{A}(k)\mathbf{L}_n = 0 \tag{4.24}$$

where $A(k) = c_1(k) + n(k)c_2^*(k)$, $\tilde{A}(k) = c_2(k) + n(k)c_1^*(k)$.

In, particular, equilibrium states of the "Hamiltonian" $H(0)$ have the form (4.22). To prove this we make volume cutoff (we replace $\mathbb{R}^d$ with a lattice). Then we can approximate $\hat{H}_0$ by a Hamiltonian of the form $\sum \epsilon_k \hat{a}_k^* \hat{a}_k$ where k runs over a discrete set. The equilibrium state of the latter Hamiltonian can be represented by density matrix $\Omega(T) = \exp(-\beta \sum \epsilon_k \hat{a}_k^* \hat{a}_k)/Z$ in the Fock space; it is easy to check that

$$\hat{a}_k \Omega(T) = e^{-\hbar \frac{\epsilon_k}{T}} \Omega(T)\hat{a}_k, \hat{a}_k^* \Omega(T) = e^{\frac{\hbar \epsilon_k}{T}} \Omega(T)\hat{a}_k^*. \tag{4.25}$$

Applying (4.17) we obtain equations for the corresponding L-functional; taking the limit we obtain for the L-functional $\mathbf{L}_T$ corresponding to the equilibrium state in infinite volume

$$c_1(k)\mathbf{L}_T = e^{-\frac{\hbar\epsilon(k)}{T}}(-\hbar c_2^*(k) + c_1(k))\mathbf{L}_T$$

$$c_2(k)\mathbf{L}_T = e^{-\frac{\hbar\epsilon(k)}{T}}(-\hbar c_1^*(k) + c_2(k))\mathbf{L}_T$$

hence the equations (4.24) are satisfied with

$$n(k) = \frac{\hbar}{e^{\frac{\hbar\epsilon(k)}{T}} - 1}. \tag{4.26}$$

4.8 Diagram Techniques in Formalism of L-functionals

Let us consider now a translation-invariant perturbation of a classical theory with quadratic Hamiltonian. Corresponding quantum Hamiltonian can be written in the form $\hat{H} = \hat{H}(0) + g\hat{V}$ where

$\hat{V} = \sum_{m,n} \int d^m k d^n l \gamma_{m,n}(k_1, ...k_m, l_1, ..., l_n)\delta(\sum \mathbf{k}_i - \sum \mathbf{l}_j)\hat{a}_1^*(k_1)...\hat{a}_m^*(k_m)$
$\hat{a}(l_1)...\hat{a}(l_n)$

where the variables are pairs $k = (\mathbf{k}, s)$, $l = (\mathbf{l}, s)$ and s stands for discrete variable.

It induces "Hamiltonian" $H = H(0) + gV$ in the space $\mathcal{L}$ and operators denoted by the same symbols H, $H(0)$, V in the dual space $\mathcal{L}^\vee$. Both in $\mathcal{L}$ and in $\mathcal{L}^\vee$ we have actions of two copies of Weyl algebra $\mathcal{W}$.

One can obtain translation-invariant stationary states of the Hamiltonian $\hat{H}$ in the formalism of L-functionals from translation-invariant stationary states of the Hamiltonian $\hat{H}(0)$ using "adiabatic dressing": $\omega = \lim_{a\to 0} S_a(0, -\infty|g)\omega_0$. In this formula we use the evolution operator $U_a(t, t_0|g)$ and the evolution operator in the interaction picture $S_a(t, t_0|g)$ defined as evolution operators corresponding to time-dependent "Hamiltonian" $H(0) + gh(a\tau)V$. (See Sect. 2.8.3 for more details.)

We will take as ω_0 a state of the form (4.22); truncated correlation functions vanish for states of this kind therefore we can expect that the state ω has cluster property. This state can be specified by conditions (4.24) that we will write in the form

$\mathbf{a}_1(k)\omega_0 = 0, \mathbf{a}_2(k)\omega_0 = 0$ where $[\mathbf{a}_i(k), \mathbf{a}_j(k'] = 0$.

Let us fix now $\alpha_0 \in \mathcal{L}^\vee$ obeying $\langle\alpha_0|\mathbf{a}_1^*(k) = 0, \langle\alpha_0|\mathbf{a}_2^*(k) = 0$ where $[\mathbf{a}_i^*(k), \mathbf{a}_j^*(k'] = 0, [\mathbf{a}_i(k), \mathbf{a}_j^*(k'] = \delta_i^j\delta(k, k')$.

We define $\alpha \in \mathcal{L}^\vee$ by the formula

$\langle\alpha| = \lim_{a\to 0}\langle\alpha_0|S_a(+\infty, 0)$

assuming that this limit exists.

Following considerations of Sect. 2.8.3 we define adiabatic GGreen functions by the formula

$$\langle\alpha_0|T\Big(\tilde{\mathrm{A}}_1(\mathbf{x}_1, t_1)...\tilde{\mathrm{A}}_m(\mathbf{x}_m, t_m)\mathrm{B}_1(\mathbf{x}_1', t_1')...\mathrm{B}_n(\mathbf{x}_n', t_n)S_a(+\infty, -\infty|g)\Big)|\omega_0\rangle \quad (4.27)$$

where $A, B \in \mathcal{W}$, $\mathrm{A}(t) = e^{-H(0)t}Ae^{H(0)t}$. If a is small and $t_i > t_{i+1}, t_j' > t_{j+1}'$ then (4.27) can be written in the form

$$\langle\alpha_0|S_a(+\infty, 0\, g)\tilde{A}_1(\mathbf{x}_1, t_1|g)...\tilde{A}_m(\mathbf{x}_m, t_m|g)B_1(\mathbf{x}_1', t_1'|g)...B_n(\mathbf{x}_n', t_n|g)S_a(0, -\infty|g)|\omega_0\rangle \quad (4.28)$$

(We used the relations $\mathrm{A}(t) = e^{-H(0)t}Ae^{H(0)t}$, $A(t|g) = e^{-H(g)t}Ae^{H(g)t} = S(0, t|g)\mathrm{A}(t)S(t, 0|g)$, $S_a(t, t_0|g) = S(t, t_0|g)$ for small a.)

Adiabatic GGreen functions tend to GGreen functions as $a \to 0$.

Let us represent the "Hamiltonian" in normal form with respect to the operators $\mathbf{a}_i^*(k), \mathbf{a}_(k)$. Using this representation we can develop diagram techniques for calculation of the normal form of the adiabatic scattering matrix, for adiabatic GGreen functions and for GGreen functions. The vertices of these diagrams are the coefficient functions of the "Hamiltonian" in normal form (when we are working with adiabatic scattering matrix or with adiabatic GGreen functions we should multiply these functions by $h(a\tau)$). The propagators are expressions $\langle\alpha_0|T(C_r(\mathbf{x}, t)C_s(\mathbf{x}', t'))|\omega_0$ where $C = \tilde{\mathrm{A}}$ or $C = \mathrm{B}$. (We use the fact that $\langle\alpha_0|D|\omega_0\rangle$ is equal to the constant term in the normal form of D,hence the propagators can be used to go from the chronological product to the normal form.)

4.8.1 Inclusive Scattering Matrix from Adiabatic Scattering Matrix in the Formalism of L-functionals

In Sect. 2.8.3 we formulated a theorem relating (in the framework of perturbation theory) the adiabatic scattering matrix in finite volume with the conventional scattering matrix. In the present section we formulate a corresponding result in the formalism of L-functionals. In this formalism the volume cutoff is not needed; the adiabatic scattering matrix in the formalism of L-functionals (denoted by $\mathbf{S}_a$) is related to inclusive scattering matrix.

In what follows we consider the most important case when $\alpha_0 = 1$ and $\omega_0 = 1$ (inclusive scattering matrix of elementary excitations of ground state—particles). In this case the normal form defined above coincides with the conventional normal form.

In the framework of perturbation theory the adiabatic scattering matrix $\mathbf{S}_a$ acts in the space of excitations of ω_0 (of bare vacuum).

Let us assume that the expression of the connected part of S_a in terms of operators c_i^*, c_i where $i = 1, 2$ (the sum of connected diagrams) has the form

$$\sum_{m,n,m',n'} \int d\mathbf{p}_i d\mathbf{q}_j d\mathbf{p}'_{i'} d\mathbf{q}'_{j'} \nu^a_{m,n,m',n'}(\mathbf{p}_1, \ldots, \mathbf{p}_m, \mathbf{q}_1, \ldots, \mathbf{q}_n, \mathbf{p}'_1, \ldots, \mathbf{p}'_{m'}, \mathbf{q}'_1, \ldots, \mathbf{q}'_{n'})\times$$

$$c_1^*(\mathbf{p}_1)\ldots c_1^*(\mathbf{p}_m)c_2^*(\mathbf{q}_1)\ldots c_2^*(\mathbf{q}_m)c_1(\mathbf{p}'_1)\ldots c_1(\mathbf{p}'_{m'})c_2(\mathbf{q}'_1)\ldots c_2(\mathbf{q}'_{n'}).$$

The coefficient functions $\nu^a_{m,n,m',n'}$ diverge as $a \to 0$, however, one can multiply these functions by such factors that the product has finite limit as $a \to 0$ and this limit is related to inclusive scattering matrix. Namely, replacing in the above expression the functions $\nu^a_{m,n,m',n'}$ by the functions

$$\nu_{m,n,m',n'} = \lim_{a\to 0} C^a_{m,n,m',n'} \nu^a_{m,n,m',n'}$$

where

$$C^a_{m,n,m',n'} = e^{i(r_a(\mathbf{p}_1)+\ldots r_a(\mathbf{p}_m))-i(r_a(\mathbf{q}_1)+\ldots r_a(\mathbf{q}_n))-i(r_a(\mathbf{p}'_1)+\ldots r_a(\mathbf{p}_{m'}))+i(r_a(\mathbf{q}'_1)+\ldots r_a(\mathbf{q}_{n'})}$$

and $r_a(\mathbf{k})$) is defined by the formula (2.60) we obtain the connected inclusive scattering matrix.

As in Sect. 2.8.3 the proof is based on the version of perturbation theory where the vertices are 1 PI diagrams and propagators are Green's functions (in this case, generalized Green's functions).

4.9 Infrared Problem in Quantum Electrodynamics

An electron in quantum electrodynamics cannot be considered as a particle in the sense of a definition accepted in this book. (This is true also for other formal definitions.) The physical reason for this is the fact that an electron generates electromagnetic field; in the language of particles this means that an electron is "dressed" by a cloud of photons. Usual scattering matrix does not make any sense for electrons: in QED the cross-section of any process involving finite number of electrons and photons is equal to zero because the collision always generates photons having very low energy (soft photons).

This means that in QED we should consider inclusive cross-sections. This was understood long ago in the paper [4]. Later the infrared problem was analyzed in numerous papers; it was shown that infrared divergences in perturbative calculations cancel in inclusive cross-sections. The formalism of L-functionals leads directly to inclusive scattering matrix; it seems that infrared divergences do not appear in this formalism. In present section we verify this fact in the situation when one can neglect the action of photons on electrons. (This is the situation considered in [4]; it is well-known that the understanding of infrared problem boils down to the analysis of this situation.) From mathematical viewpoint the infrared problem appears due to the fact that the state of electron dressed by photons does not belong to Fock space. The correct Hilbert space for dressed electron was described in [16]. As was noticed above L-functionals describe states in all representations of CCR, hence the problem of finding the correct Hilbert space does not appear. Our calculations agree with results of [16].

Let us show that in the formalism of L-functionals infrared divergences do not appear for time-dependent Hamiltonian

$$\hat{H} = \hat{H}_0 + \hat{V} = \hbar \int dk\epsilon(\mathbf{k})a^*(\mathbf{k}) \cdot a(\mathbf{k})$$
$$+\hbar \int \frac{d\mathbf{k}}{\sqrt{2\epsilon(\mathbf{k})}}(j(\mathbf{k},t) \cdot a^*(\mathbf{k}) + j^*(\mathbf{k},t) \cdot a(\mathbf{k})).$$

Here $a(\mathbf{k})$ is a vector potential of electromagnetic field with components $a_\mu(\mathbf{k}), \mu = 0, .., 3$ satisfying the Lorenz gauge condition $k^\mu a_\mu(\mathbf{k}) = 0$. The scalar product of two 4-vectors has the form $p \cdot k = \mathbf{pk} - p_0 k_0$. We use the notation $\epsilon(\mathbf{k}) = |\mathbf{k}|$.

We suppose that the $j(\mathbf{k}, t) = j^\mu(\mathbf{k}, t)$ is a numerical function (Fourier transform of divergence-free current). For example, we can consider an electron moving in a potential field and interacting with a quantized electromagnetic field (we neglect the action of electromagnetic field on the electron).

The equation of motion for L-functional corresponding to this Hamiltonian has the form, $d\mathbf{L}/dt = H\mathbf{L}$ where $H = H_0 + V(t)$ and

$$H_0 = \hbar \int d\mathbf{k}\epsilon(\mathbf{k})(c_1^*(\mathbf{k}) \cdot c_1(\mathbf{k}) - c_2^*(\mathbf{k}) \cdot c_2(\mathbf{k})),$$

$$V(t) = \hbar \int \frac{d\mathbf{k}}{\sqrt{2\epsilon(\mathbf{k})}}(j(\mathbf{k},t) \cdot c_1^*(\mathbf{k}) + j^*(\mathbf{k},t) \cdot c_2^*(\mathbf{k}))$$

The corresponding evolution operator will be denoted by $U(t)$. Working in the interaction picture (= making the change of variables $L_I(t) = e^{iH_0 t/\hbar}L(t)$) we reduce the calculation of $U(t)$ to the calculation of the operator $S(t) = e^{-iH_0 t/\hbar}U(t)e^{iH_0 t/\hbar}$ obeying $i\frac{dS}{dt} = \mathrm{V}(t)S$ where

$\mathrm{V}(t) = \int \frac{d\mathbf{k}}{\sqrt{2\epsilon(\mathbf{k})}}(e^{i\epsilon(\mathbf{k})t}j(\mathbf{k},t) \cdot c_1^*(\mathbf{k}) + e^{-i\epsilon(\mathbf{k})t}j^*(\mathbf{k},t) \cdot c_2^*(\mathbf{k}))$
We obtain $S = e^{M_1(t)+M_2(t)}$

where we use the notation

$M_i(t) = \int \frac{d\mathbf{k}}{\sqrt{2\epsilon(\mathbf{k})}}(M_i(\mathbf{k}, t) \cdot c_i^*(\mathbf{k}))$.

By comparing this expression with equation of motion we get

$\dot{M}_1(\mathbf{k}, t) = j(\mathbf{k}, t)e^{i\epsilon(\mathbf{k})t}$;
$\dot{M}_2(\mathbf{k}, t) = j^*(\mathbf{k}, t)e^{-i\epsilon(\mathbf{k})t}$;

Using the definition of operators c_i^* we can write the expression for the solution of the equation of motion in the formalism of L-functionals in the following form

$$L(\alpha^*, \alpha, t) = \exp\Big(\int_{t_0}^{t} d\tau \int \frac{d\mathbf{k}}{\sqrt{2\epsilon(\mathbf{k})}}(e^{i\epsilon(\mathbf{k})\tau} j(\mathbf{k}, \tau) \cdot \alpha^*(\mathbf{k}) + e^{-i\epsilon(\mathbf{k})\tau} j^*(\mathbf{k}, \tau) \cdot \alpha(\mathbf{k}))\Big)L(\alpha^*, \alpha, t_0) \tag{4.29}$$

The formula for the solution can be rewritten in the form

$$L(\alpha^*, \alpha, t) = \exp\Big(\int d\mathbf{k}\sqrt{2\epsilon(\mathbf{k})}(e^{-i\epsilon(\mathbf{k})t}\mathcal{A}(\mathbf{k}, t)\dot{\alpha}^*(\mathbf{k}) + e^{i\epsilon(\mathbf{k})t}\mathcal{A}^*(\mathbf{k}, t) \cdot \alpha(\mathbf{k})\Big)L(\alpha^*, \alpha, t_0)$$

where

$$\mathcal{A}^{\mu}(\mathbf{k}, t) = \frac{1}{2\epsilon(\mathbf{k})(2\pi)^{\frac{3}{2}}} \int_{t_0}^{t} d\tau (e^{i\epsilon(\mathbf{k})(\tau - t)} j^{\mu}(\mathbf{k}, \tau).$$

Here $\mathcal{A}^{\mu}(\mathbf{k}, t)$ is the expectation value of electromagnetic potential. Indeed, it is easy to check that

$$\mathcal{A}_{\mu}(\mathbf{k}, t) = \frac{e^{i\epsilon(\mathbf{k})t}}{\sqrt{2\epsilon(\mathbf{k})}}\tilde{b}_{\mu}(\mathbf{k})L(\alpha, \alpha^*, t)|_{\alpha=0}$$
$$= \frac{e^{i\epsilon(\mathbf{k})t}}{\sqrt{2\epsilon(\mathbf{k})}}\frac{\delta}{\delta\alpha_{\mu}^*}L(\alpha, \alpha^*, t)|_{\alpha=0}$$

Let us now turn to calculating the inclusive cross-section of the photon emission. For this purpose, we calculate the expectation value of the operator

$$\rho(\mathbf{k}) = \sum_{i=\pm}(\varepsilon_i^* \cdot a^*(\mathbf{k}))(\varepsilon_i \cdot a(\mathbf{k})),$$

where ε_i are polarizations of outgoing photons. We get

$$\begin{aligned} dN(\mathbf{k}) &= \langle \rho(\mathbf{k}) \rangle d\mathbf{k} \\ &= \left(\sum_{i=\pm} \varepsilon_i \frac{\delta}{\delta \alpha(\mathbf{k})} \varepsilon_i^* \frac{\delta}{\delta \alpha^*(\mathbf{k})} L(\alpha, \alpha^*, t)|_{\alpha=0} \right) 2\epsilon(\mathbf{k}) d\mathbf{k} \\ &= \mathcal{A}(\mathbf{k}, t) \cdot \mathcal{A}^*(\mathbf{k}, t) 2\epsilon(\mathbf{k}) d\mathbf{k} \end{aligned}$$

If we are interested in the inclusive cross-section of emission of n photons with momenta $k_1, ..., k_n$ then similar calculations lead to the following formula:

$$\begin{aligned} dN(\mathbf{k}_1, ..., \mathbf{k}_n) &= \langle \rho(\mathbf{k}_1, ..., \mathbf{k}_n) \rangle \prod_{i=1}^{n} d\mathbf{k}_i \\ &= \prod_{i=1}^{n} \mathcal{A}(\mathbf{k}_i, t) \cdot \mathcal{A}^*(\mathbf{k}_i, t) 2\epsilon(\mathbf{k}_i) d\mathbf{k}. \end{aligned}$$

As a simplest example we consider a free electron with momentum $\mathbf{p}$. Then the current $j^\mu(\mathbf{k}, t)$ has the form $\mathrm{e}\frac{p^\mu}{p_0} e^{i\omega_p t}$ where $\omega_p(\mathbf{k}) = \frac{\mathbf{kp}}{p_0}$.

Applying the formula (4.29) we obtain the L-functional describing the cloud of photons around the electron.

Another example is the emission of photons by an electron changing its momentum $\mathbf{p}$ to $\mathbf{p}'$. Then we can take the current in the following form

$$j^\mu(\mathbf{k}, t) = \mathrm{e}\theta(-t) \frac{p^\mu}{p_0} e^{i\omega_p t} + \mathrm{e}\theta(t) \frac{p'^\mu}{p'_0} e^{i\omega_{p'} t}. \tag{4.30}$$

Calculating the inclusive cross-section of the photon emission in this process we obtain

$$dN(\mathbf{k}) = \frac{\mathrm{e}^2}{2\epsilon(\mathbf{k})} \left| \frac{p'}{p' \cdot k} - \frac{p}{p \cdot k} \right|^2 \frac{d\mathbf{k}}{(2\pi)^3} \tag{4.31}$$

We considered photons interacting with classical current j^μ. Very similar calculations can be performed for gravitons interacting with classical gravitational field (for example the field generated by a collision of two black holes; this is the situation considered, for example, in [5]).

We considered quantum electrodynamics in the case when the action of photons on electrons can be neglected. In general case when photons interact with other particles we also can apply the formalism of L-functionals. We denoted the arguments of L-functionals for photons by $\bar{f}, f$, corresponding multiplication operators by c_1^*, c_2^* and corresponding variational derivative by c_1, c_2, now we have new arguments corresponding to other particles, new multiplication operators γ_1^*, γ_2^* and new variational derivatives denoted by γ_1, γ_2. (Notice that operators or, more precisely, generalized operator functions, c_i^*, c_i depend on momentum $\mathbf{p}$ and discrete variable taking two values (photon polarization), operators γ_i^*, γ_i also depend on $\mathbf{p}$ and discrete variable.)

The adiabatic scattering matrix $\mathbf{S}_a$ is well-defined in this formalism also the case when photons interact with massive fermions or bosons. It acts in the space of excitations of the bare vacuum state ω_0.

Let us express the connected part of $\mathbf{S}_a$ in normal form with respect to operators $\gamma_i^*, c_i^*, \gamma_i, c_i$. The coefficient functions of this expression diverge as $a \to 0$.

Conjecture Multiplying these coefficient functions by some numerical factors depending on a and taking the limit $a \to 0$ we obtain a finite expression that can be interpreted as inclusive scattering matrix.

Correction to: Quantum Mechanics and Quantum Field Theory from Algebraic and Geometric Viewpoints

Correction to:
A. Schwarz, *Quantum Mechanics and Quantum Field Theory from Algebraic and Geometric Viewpoints*, SpringerBriefs in Physics, https://doi.org/10.1007/978-3-031-67915-5

The book was inadvertently published without updated online "Abstract" text for "Chapters 01, 02, 03, and 04". This has been included in the online version.

The updated version of this book can be found at
https://doi.org/10.1007/978-3-031-67915-5

A. Schwarz, *Quantum Mechanics and Quantum Field Theory from Algebraic and Geometric Viewpoints*,
SpringerBriefs in Physics, https://doi.org/10.1007/978-3-031-67915-5_5

Appendix
Background

All sections of Appendix contain material reprinted from the review [10].

A.1 Notations, Conventions, Definitions

Talking about vector spaces we always assume vector spaces over $\mathbb{R}$ or over $\mathbb{C}$. (If the field is not specified we have in mind a complex vector space.)

An algebra is a vector space over $\mathbb{R}$ or over $\mathbb{C}$ equipped with an operation of multiplication satisfying the distributivity axiom $(a+b)c = ac + bc$, $c(a+b) = ca + cb$, where a, b are elements of the algebra, c is an element of the algebra or a number.

An algebra is unital if there exists an element 1 obeying $1 \cdot a = a \cdot 1 = a$.

An algebra is associative if $(ab)c = a(bc)$.

A $*$-algebra is an associative algebra with antilinear involution * obeying $a^{**} = a$, and $(ab)^* = b^* a^*$. A homomorphism or automorphism of $*$-algebra should agree with involution. We use notations $\mathcal{A}$ for $*$-algebra and $Aut(\mathcal{A})$ for its group of automorphisms.

The algebra $End(\mathcal{H})$ of bounded linear operators in Hilbert space $\mathcal{H}$ is a $*$-algebra with respect to the involution $A \to A^*$ where A^* is the operator adjoint to A. (Notice that for bounded operators "adjoint" is the same as "Hermitian conjugate", but for unbounded operators these notions are different. Operator B is Hermitian conjugate to A if $\langle Ax, y\rangle = \langle x, By\rangle$ for $x \in D_A$, $y \in D_B$ in corresponding domains D_A, D_B. Adjoint operator A^* can be defined as Hermitian conjugate with maximal domain. We disregard this distinction).

We say that a self-adjoint operator A is positive definite if $\langle x, Ax\rangle \geq 0$ for any element $x \in D_A$. (In more standard terminology such operators are called positive semi-definite.) A positive definite operator with eigenvalues λ_n belongs to the trace class if the series $\sum \lambda_n$ converges; an arbitrary operator A belongs to the trace class

A. Schwarz, *Quantum Mechanics and Quantum Field Theory from Algebraic and Geometric Viewpoints*, SpringerBriefs in Physics,
https://doi.org/10.1007/978-3-031-67915-5

if the operator $\sqrt{A^*A}$ is in this class. Operators belonging to the trace class have well-defined trace.

A Lie algebra is an algebra with an operation $[a, b]$ obeying $[a, b] = -[b, a]$ and $[[ab], c] + [[b, c], a] + [[c, a], b] = 0$ (Jacobi identity).

A topological group, vector space, algebra,... is a group, vector space, algebra,...equipped with topology in such a way that all operations (multiplication, addition, multiplication by a number) are continuous.

We always assume that a map (=mapping) of topological spaces (or homomorphism of topological groups, etc) is continuous.

Notice that sometimes it is convenient to consider topological algebras where the operation of multiplication is defined only on a dense subset. For example, we can consider the Lie algebra of (not necessarily bounded) self-adjoint operators in Hilbert space. In this algebra the commutator is not necessarily well-defined but in appropriate topology it is defined on a dense subset.

A derivation of an algebra is a linear operator D obeying Leibniz rule $D(ab) = Da \cdot b + a\dot{D}b$. A commutator of derivations is a derivation, hence we can talk about the Lie algebra of derivations.

If we are dealing with a topological algebra we can consider derivations defined on a dense subset. Their commutator is not necessarily well-defined, but still, we can regard the set of such derivations as topological Lie algebra that can be considered as Lie algebra of the automorphism group of the algebra.

We say that the derivation D is an infinitesimal automorphism if it can be considered as a tangent vector of a one-parameter subgroup of the automorphism group, i.e. there exists a solution of the equation $dU/dt = DU(t)$ with initial condition $U(0) = 1$. The solution can be written as $\exp(Dt)$; this can be regarded as the definition of the operator exponent.

The set of infinitesimal automorphisms can be regarded as topological Lie algebra, which also can be interpreted as Lie algebra of the group of automorphisms.

More generally if we have a topological group we can define a Lie algebra of this group either starting with the set of tangent vectors to the curves in the group at the unit element or starting with the set of tangent vectors to one-parameter subgroups.

Notice that the above definitions and statements are not rigorous. For example, we did not specify the topology in the group of automorphisms, we did not give a definition of tangent vector, etc. One should remember that one can give several reasonable definitions of these and related notions. We always disregard these subtleties.

We denote by $\mathcal{S} = \mathcal{S}(\mathbb{R}^n)$ the space of smooth fast decreasing functions on $\mathbb{R}^n$ (Schwartz space). More precisely, this space consists of smooth functions such that

$$\sup_{x \in \mathbb{R}^n} |x_1^{\alpha_1} ... x_n^{\alpha_n} \partial_1^{\beta_1} ... \partial_n^{\beta_n} f(x)| \tag{A.1}$$

is finite for any choice of non-negative integers α_i, β_j.

The expressions (A.1) can be regarded as seminorms specifying the topology in $\mathcal{S}$.

We consider generalized functions on $\mathbb{R}^n$ (distributions) as continuous linear functionals on $\mathcal{S}$.

More generally, generalized functions are defined as linear functionals on some topological vector space of functions on $\mathbb{R}^n$ (on the space of test functions); one represents such a functional f as formal integral: $f(\phi) = \int dx f(x)\phi(x)$.

In the above definitions test functions (hence generalized functions) can be regarded as functions taking values in a space $\mathbb{C}^r$ (as vector-valued functions).

The elementary space $\mathfrak{h}$ can be considered as a space of smooth fast decreasing functions on $\mathbb{R}^d$ taking values in $\mathbb{C}^r$.

A spatial translation by $\mathbf{a} \in \mathbb{R}^n$ (=spatial shift) is denoted by $T_{\mathbf{a}}$. In coordinate representation, it acts on an element f of $\mathfrak{h}$ as the shift of the argument $\mathbf{x}$ by $\mathbf{a}$,

$$T_{\mathbf{a}} f(\mathbf{x}) = f(\mathbf{x} + \mathbf{a}),$$

in momentum representation, it acts as multiplication by the function $e^{i\mathbf{ak}}$. (Coordinate and momentum representations are related by Fourier transform.)

Temporal translations (=time translations = time shifts) are denoted by T_τ. They should commute with spatial translations; it follows that in momentum representation they act on the elements of $\mathfrak{h}$ by the formula $(T_\tau f)(\mathbf{k}) = e^{-i\tau E(\mathbf{k})} f(\mathbf{k})$, where $E(\mathbf{k})$ is an $r \times r$ Hermitian matrix.

The notations $T_{\mathbf{a}}$ and T_τ are used for spatial and temporal translations not only in the elementary space but also in other situations. Together the space and time translations generate a commutative group denoted by $\mathcal{T}$.

We always assume summation over repeated indices.

A.2 Convex Sets

This appendix contains basic information about convex sets, which is used throughout the book.

A convex set C is a subset of vector space that, together with every two points, contains a segment connecting these points. In a convex set one can consider a mixture of points of the set. If we take some points of the set and ascribe a non-negative number to each point such that the sum of numbers equals one, then the sums of points $e_i \in C$ with coefficients $p_i \geq 0, \sum p_i = 1$ will also belong to the convex set. The sum $\sum p_i e_i \in C$ is called the mixture of points e_i with probabilities p_i. One can consider the numbers p_i as weights, and then this sum will represent the center of gravity. Another important notion is that of the extreme point of a convex set. An extreme point is a point that does not lie inside any segment with ends belonging to the set. The extreme points of a polyhedron are vertices. For a ball, the extreme points lie on the boundary sphere.

We will only consider convex subsets of a complete topological vector spaces. We will assume that all convex sets we consider are closed; then one can consider a mixture not only of a finite number of points but also a mixture of a countable number of points. In the latter case, the sum representing the mixture might be infinite. Convex envelope of a subset E of a complete topological vector space is the smallest convex subset of the space containing E.

A convex cone is defined as a closed convex set that is invariant with respect to dilations $x \to \lambda x$ for any $\lambda \in \mathbb{R}_+$. Notice, that this definition of a convex cone is not quite standard, usually one imposes some additional conditions. We use the notation C for convex cones.

It is also possible to consider an uncountable mixture of points of any subset of a convex set if the subset is equipped with a probability measure. For example, let ω denote points on a sphere S bounding a ball B and let μ be a measure on S. Suppose that there is a probability measure on the sphere with probability density ρ with respect to μ. Then we can consider uncountable mixtures of points of the sphere: just replace the sum with the integral: $\int_S \omega\rho(\omega)d\mu(\omega) \in B$.

A *compact* convex set C is completely determined by its extreme points. Namely, one has the following theorem (Krein-Milman): if C is a compact convex subset of a locally convex topological vector space, then C is the closed convex envelope of its extreme points. In particular, such a set has extreme points. Under the same set of assumptions, one can say that every point of C is a mixture of extreme points (Cloquet-Bishop-de Leeuw theorem).

A.3 Asymptotic Behavior of Solutions of Linear Equations

Let us consider solutions of translation-invariant linear equation

$$\frac{\partial f}{\partial t} = Af. \tag{A.2}$$

We assume that for fixed t the function $f(\mathbf{x}, t)$ is defined on $\mathbb{R}^d$, it takes values in $\mathbb{C}^r$. It follows from translation invariance that after Fourier transform with respect to $\mathbf{x}$ (i.e. in the momentum representation) the operator A can be regarded as an operator of multiplication of $f(\mathbf{p})$ considered as a column vector by an $r \times r$ matrix $A(\mathbf{p})$. If the operator A is local (represented as a polynomial of derivatives) then the matrix $A(\mathbf{p})$ is a polynomial. We do not assume locality, but we suppose that $A(\mathbf{p})$ is a smooth function of $\mathbf{p}$) (then we can say that A is quasi-local).

The solution to (A.2) in momentum representation has the form $e^{tA(\mathbf{p})} f(\mathbf{p})$.

Let us assume that the matrix $A(\mathbf{p})$ is diagonalizable and has purely imaginary eigenvalues. Then the solution $f \equiv 0$ is stable. (This means that for an appropriate definition of the norm, the evolution operators T_τ are uniformly bounded: $|T_\tau|| \leq C$ for all τ. In particular, if f is small at some moment it remains small as $\tau \to \pm\infty$.).

Let's consider the solution to the Eq. (A.2) in coordinate representation

$$f(\mathbf{x}, t) = \int d\mathbf{p} e^{i\mathbf{p}\mathbf{x}} e^{tA(\mathbf{p})} f(\mathbf{p}). \tag{A.3}$$

The same formula describes the behavior of a test function (of an element of an elementary space $\mathfrak{h}$) in the coordinate representation if we take $E(\mathbf{p}) = -iA(\mathbf{p})$.

We are interested in the behavior of the function (A.3) as $t \to \pm\infty$. Diagonalizing the matrix $A(\mathbf{p})$ we can reduce this problem to the case when $r = 1$. (This statement is not quite rigorous, because smooth dependence of $A(\mathbf{p})$ of its argument does not imply that eigenvalues are smooth functions of $\mathbf{p}$; one should avoid diagonalization in rigorous treatment.) We assume that $A(\mathbf{p}) = -i\epsilon(\mathbf{p})$ following the notations of the preceding section. To analyze the behavior of the function

$$(T_\tau f)(\mathbf{x}) = f(\mathbf{x}, \tau) = \int d\mathbf{p} e^{i\mathbf{p}\mathbf{x} - i\tau\epsilon(\mathbf{p})} f(\mathbf{p}) \tag{A.4}$$

we notice that for large $|\tau|$ the phase is large and we can use the stationary phase method. It leads to equations:

$$\frac{\mathbf{x}}{\tau} = \nabla\epsilon(\mathbf{p}). \tag{A.5}$$

Clearly, we must consider only the situation when $\mathbf{p} \in \operatorname{supp} f$. (Recall that $\operatorname{supp} f$-the support of the function f in momentum space -is defined as the closure of the set of points where $f(\mathbf{p}) \neq 0$. We assume that the set $\operatorname{supp} f$ is compact.)

Let us define the set U as a neighborhood of the set of points $\mathbf{x}$ where the condition (A.5) with $\mathbf{p} \in \operatorname{supp} f$ and $\tau = 1$ is satisfied. Outside the set τU the Eq. (A.5) has no solution, so the function $(T_\tau f)(\mathbf{x})$ is very small at $\mathbf{x} \notin \tau U$.

We say that τU is the essential support of the function f in coordinate representation for large $|\tau|$. One can prove that for $\mathbf{x} \notin \tau U$ we have

$$|(T_\tau f)(\mathbf{x})| < C_n(1 + |\mathbf{x}|^2 + \tau^2)^{-n} \tag{A.6}$$

for any integer n. The proof can be based on a generalization of the Riemann–Lebesgue lemma. Recall that it follows from this lemma that the function $\int dk g(k) e^{ih(k)t}$ where $g(k)$ is a smooth function having compact support and $h(k)$ is a linear function tends to zero faster than any power of t. (To prove this statement we integrate by parts many times.) This statement can be easily generalized to the case when the function $h(k)$ is not linear but is smooth and does not have stationary points. Then locally this function can be made linear by means of a change of variables. Using a partition of unity (a representation of unity as a finite sum of functions $g_a(k)$ that do not vanish only on small sets U_a covering the support of $g(k)$) and linearizing $h(k)$ on the sets U_a we obtain the generalization we need. The Riemann–Lebesgue lemma and its generalization can be proven also in the multi-dimensional case; this allows us to verify the above estimate.

When we apply the stationary phase method to the calculation of the integral (A.4) we obtain a factor $(\det(\tau \mathtt{Hess}))^{-\frac{1}{2}} = \tau^{-\frac{d}{2}}(\det \mathtt{Hess})^{-\frac{1}{2}}$ where Hess denotes the matrix of second derivatives. This remark allows us to conjecture that

$$|f(\mathbf{x}, \tau)| \leq C\tau^{\frac{-d}{2}}. \tag{A.7}$$

The estimate (A.7) can be used to analyze the problem of the existence of a solution of the non-linear Eq. (2.1) with given asymptotical behavior as $t \to -\infty$ [34].

References

1. Albrecht A (1992) Investigating decoherence in a simple system. Phys Rev D 46(12):5504
2. Araki H, Haag R (1967) Collision cross sections in terms of local observables. Commun Math Phys 4:77–91
3. Berezin FA (2012) The method of second quantization, vol 24. Elsevier
4. Bloch F, Nordsieck A (1937) Note on the radiation field of the electron. Phys Rev 52(2):54
5. Caron-Huot S, Giroux M, Hannesdottir HS, Mizera S (2024) What can be measured asymptotically? J High Energy Phys 2024(1):1–63
6. Chou KC, Su ZB, Hao BL, Yu L (1985) Equilibrium and nonequilibrium formalisms made unified. Phys Rep 118(1–2):1–131
7. Chu H, Umezawa H (1994) A unified formalism of thermal quantum field theory. Int J Mod Phys A 9(14):2363–2409
8. Faddeev LD, Korepin VE (1978) Quantum theory of solitons. Phys Rep 42(1):1–87
9. Fateev V, Schwarz A (1973) On axiomatic scattering theory. Teoreticheskaya i Matematicheskaya Fizika 14(2):152–69
10. Frolov I, Schwarz A (2023) Quantum mechanics and quantum field theory: algebraic and geometric approaches. Universe 9(7):337
11. Frolov I, Schwarz A (in preparation)
12. Hunziker W, Sigal IM (2000) The quantum n-body problem. J Math Phys 41(6):3448–3510
13. Jordan P (1933) Ueber die multiplikation quantenmechanischer groessen. Zeitschrift für Physik 80(5):285–291
14. Jordan P, von Neumann J, Wigner EP (1993) On an algebraic generalization of the quantum mechanical formalism. In: The collected works of Eugene Paul Wigner: part a: the scientific papers. Berlin, Heidelberg: Springer, Berlin, Heidelberg, pp 298–333
15. Kontsevich M (2003) Deformation quantization of Poisson manifolds. Lett Math Phys 66:157–216
16. Kulish PP, Faddeev LD (2016) Asymptotic conditions and infrared divergences in quantum electrodynamics. In: Fifty years of mathematical physics: selected works of Ludwig Faddeev, pp 144–156
17. Likhachev VN, Tyupkin YS, Schwarz AS (1972) Adiabatic theorem in quantum field theory. Teoreticheskaya i Matematicheskaya Fizika 10(1):63–84
18. Liu B, Soffer A (2023) The large time asymptotic solutions of nonlinear Schrödinger type equations. Appl Numer Math

A. Schwarz, *Quantum Mechanics and Quantum Field Theory from Algebraic and Geometric Viewpoints*, SpringerBriefs in Physics,
https://doi.org/10.1007/978-3-031-67915-5

19. Von Neumann J (2018) Mathematical foundations of quantum mechanics: new edition, vol 53. Princeton university press
20. Paz JP, Zurek WH (1999) Quantum limit of decoherence: environment induced super-selection of energy eigenstates. Phys Rev Lett 82(26):5181
21. Shvarts AS (1967) New formulation of quantum theory. Dokl Akad Nauk SSSR 173:793
22. Schwarz A, Tyupkin YS (1987) Measurement theory and the Schroedinger equation. In: Quantum field theory and quantum statistics: essays in honour of the sixtieth birthday of ES Fradkin. V. 1
23. Schwarz A (2010) Space and time from translation symmetry. J Math Phys 51(1):015201
24. Schwarz A (2020) Inclusive scattering matrix and scattering of quasiparticles. Nucl Phys B 950:114869
25. Schwarz A (2020) Mathematical foundations of quantum field theory. World Scientific
26. Schwarz A (2020) Geometric approach to quantum theory. SIGMA Symmetry Integr Geom Methods Appl 16:020
27. Schwarz A (2021) Geometric and algebraic approaches to quantum theory. Nucl Phys B 973:115601. arXiv:2102.09176
28. Schwarz A (2021) Scattering in algebraic approach to quantum theory. Associative algebras. arXiv:2107.08553
29. Schwarz A (2021) Scattering in geometric approach to quantum theory. Universe 8(12): 663. arXiv:2107.08557
30. Schwarz A Scattering in algebraic approach to quantum theory. Jordan algebras. Universe 9(4):173
31. Schwarz A (in preparation)
32. Segal I (1976) Space-time decay for solutions of wave equations. Adv Math 22(3):305–311
33. Soffer A (2006) Soliton dynamics and scattering. Int Congr Math 3:459–471
34. Strauss WA (1974) Nonlinear scattering theory. In: scattering theory in mathematical physics: proceedings of the NATO advanced study institute held at Denver, Colo., USA, June 11–29, 1973. Springer, Dordrecht, Netherlands, pp 53–78
35. Tao T (2009) Why are solitons stable? Bull Am Math Soc 46(1):1–33
36. Tyupkin YS (1973) On the adiabatic definition of the S matrix in the formalism of L functionals. Teoreticheskaya i Matematicheskaya Fizika 16(2):169–177
37. Tyupkin YS, Fateev VA, Shvarts AS (1975) Classical limit of the S matrix in quantum field theory. Sov Phys Dokl 20:194
38. Weinberg S (1989) Testing quantum mechanics. Ann Phys 194(2):336–386
39. Wigner E (1939) On unitary representations of the inhomogeneous Lorentz group. Ann Math :149–204